VOLUME ONE HUNDRED AND EIGHTY NINE

METHODS IN CELL BIOLOGY

Immuno-Oncology and Immunotherapy - Part A

SERIES EDITOR

Lorenzo Galluzzi
Fox Chase Cancer Center,
Philadelphia, PA, United States

VOLUME ONE HUNDRED AND EIGHTY NINE

METHODS IN CELL BIOLOGY

Immuno-Oncology and Immunotherapy - Part A

Edited by

NORMA BLOY
Weill Cornell Medicine, New York, NY, United States

MAUD CHARPENTIER
Weill Cornell Medicine, New York, NY, United States

LORENZO GALLUZZI
Fox Chase Cancer Center, Philadelphia, PA, United States

ELSEVIER

ACADEMIC PRESS
An imprint of Elsevier

Academic Press is an imprint of Elsevier
125 London Wall, London EC2Y 5AS, United Kingdom
525 B Street, Suite 1650, San Diego, CA 92101, United States
50 Hampshire Street, 5th Floor, Cambridge, MA 02139, United States

First edition 2024

ISBN: 978-0-443-29622-2
ISSN: 0091-679X

For information on all Academic Press publications
visit our website at https://www.elsevier.com/books-and-journals

Publisher: Zoe Kruze
Editorial Project Manager: Devwart Chauhan
Production Project Manager: A. Maria Shalini
Cover Designer: Arumugam Kothandan

Typeset by STRAIVE, India

Contents

Contributors

Olivier Adotévi
INSERM, EFS BFC, UMR 1098, RIGHT, Greffon-Hôte-Tumeur Interactions/Ingénierie Cellulaire et Génique, Université de Franche-Comté; Department of Medical Oncology, University Hospital of Besançon, Besançon, France

Ainhoa Amarilla-Irusta
Immunopathology Group, Biobizkaia Health Research Institute, Barakaldo, Spain

Laura Amo
Immunopathology Group, Biobizkaia Health Research Institute, Barakaldo; Ikerbasque, Basque Foundation for Science, Bilbao, Spain

Alberto Anel
University of Zaragoza/Aragón Health Research Institute, Biochemistry and Molecular and Cell Biology, Zaragoza, Spain

Fernando Aranda
Program of Immunology and Immunotherapy, Cima Universidad de Navarra; Navarra Institute for Health Research (IDISNA), Pamplona, Spain

Leire Arrizabalaga
Program of Immunology and Immunotherapy, Cima Universidad de Navarra; Navarra Institute for Health Research (IDISNA), Pamplona, Spain

Jérémy Baude
Department of Radiation Oncology; Preclinical Radiation Therapy and Radiobiology Unit, Centre Georges-François Leclerc, Dijon, France

Virginia Belsúe
Program of Immunology and Immunotherapy, Cima Universidad de Navarra; Navarra Institute for Health Research (IDISNA), Pamplona, Spain

Manuel Beltrán-Visiedo
Department of Radiation Oncology, Weill Cornell Medical College, New York, NY, United States

Pedro Berraondo
Program of Immunology and Immunotherapy, Cima Universidad de Navarra; Navarra Institute for Health Research (IDISNA), Pamplona; Centro de Investigación Biomédica en Red de Cáncer (CIBERONC), Madrid, Spain

Norma Bloy
Department of Radiation Oncology, Weill Cornell Medical College, New York, NY, United States

Francisco Borrego
Immunopathology Group, Biobizkaia Health Research Institute, Barakaldo; Ikerbasque, Basque Foundation for Science, Bilbao, Spain

Laura Boulerot
INSERM, EFS BFC, UMR 1098, RIGHT, Greffon-Hôte-Tumeur Interactions/Ingénierie Cellulaire et Génique, Université de Franche-Comté, Besançon, France

Jihane Boustani
INSERM, EFS BFC, UMR 1098, RIGHT, Greffon-Hôte-Tumeur Interactions/Ingénierie Cellulaire et Génique, Université de Franche-Comté; Department of Radiation Oncology, University Hospital of Besançon, Besançon, France

Chiara Camisaschi
IRCCS Humanitas Research Hospital, Rozzano, Milan, Italy

Ginevra Campia
Department of Radiation Oncology, Weill Cornell Medical College, New York, NY, United States

Julie Coupey
Université de Caen Normandie, CNRS, Normandie Université, ISTCT UMR6030, GIP Cyceron, Caen, France

Gabriele De Simone
IRCCS Humanitas Research Hospital, Rozzano, Milan, Italy

Anna Fialová
Sotio, Prague, Czech Republic

Lisa Froidurot
Preclinical Radiation Therapy and Radiobiology Unit, Centre Georges-François Leclerc, Dijon, France

Claudia Galassi
Department of Radiation Oncology, Weill Cornell Medical College, New York, NY, United States

Giovanni Galletti
Department of Molecular Biology, School of Biological Sciences, University of California, San Diego, San Diego, CA, United States

Lorenzo Galluzzi
Department of Radiation Oncology, Weill Cornell Medical College; Sandra and Edward Meyer Cancer Center; Caryl and Israel Englander Institute for Precision Medicine, New York, NY, United States

Ozgu Gumustekin
Tumor Stroma Interactions Group, Department of Cancer Research, Luxembourg Institute of Health, Luxembourg, Luxembourg

Erika S. Hue
LABÉO; Normandie Université, UNICAEN BIOTARGEN, Saint-Contest, France

Floriana Iacobone
Department of Oncology and Molecular Medicine, Istituto Superiore di Sanità (ISS), Rome, Italy

Nelia Jiménez-Aldúan
Apoptosis, Immunity & Cancer Group, IIS Aragón, University of Zaragoza, Zaragoza, Spain

Oliver Kepp
Equipe Labellisée Par La Ligue Contre Le Cancer, Université de Paris, Sorbonne Université, Centre de Recherche des Cordeliers, Institut Universitaire de France, Paris; Metabolomics and Cell Biology Platforms, Gustave Roussy, Université Paris Saclay, Villejuif, France

Vanessa Klapp
Tumor Stroma Interactions Group, Department of Cancer Research, Luxembourg Institute of Health, Luxembourg; Faculty of Science, Technology and Medicine, University of Luxembourg, Esch-sur-Alzette, Luxembourg

Vladimír Koucký
Sotio; Department of Otorhinolaryngology and Head and Neck Surgery, First Medical Faculty, Motol University Hospital, Prague, Czech Republic

Guido Kroemer
Equipe Labellisée Par La Ligue Contre Le Cancer, Université de Paris, Sorbonne Université, Centre de Recherche des Cordeliers, Institut Universitaire de France; Pôle de Biologie, Hôpital européen Georges Pompidou, AP-HP, Paris; Metabolomics and Cell Biology Platforms, Gustave Roussy, Université Paris Saclay, Villejuif, France

Riad Ladjohounlou
Preclinical Radiation Therapy and Radiobiology Unit, Centre Georges-François Leclerc, Dijon, France

Anne Largeot
Tumor Stroma Interactions Group, Department of Cancer Research, Luxembourg Institute of Health, Luxembourg, Luxembourg

Marine M. Leblond
Department of Oncology UNIL CHUV; Ludwig Institute for Cancer Research, University of Lausanne, Lausanne, Switzerland

Benoît Lecoester
INSERM, EFS BFC, UMR 1098, RIGHT, Greffon-Hôte-Tumeur Interactions/Ingénierie Cellulaire et Génique, Université de Franche-Comté, Besançon, France

Emeric Limagne
University of Burgundy; Cancer Biology Transfer Platform, Centre Georges-François Leclerc; Genetic and Immunology Medical Institute; INSERM LNC-UMR1231, Dijon, France

Peng Liu
Equipe Labellisée Par La Ligue Contre Le Cancer, Université de Paris, Sorbonne Université, Centre de Recherche des Cordeliers, Institut Universitaire de France, Paris; Metabolomics and Cell Biology Platforms, Gustave Roussy, Université Paris Saclay, Villejuif, France

Ainara Lopez-Pardo
Immunopathology Group, Biobizkaia Health Research Institute, Barakaldo, Spain

Enrico Lugli
IRCCS Humanitas Research Hospital, Rozzano, Milan, Italy

Iole Macchia
Department of Oncology and Molecular Medicine, Istituto Superiore di Sanità (ISS), Rome, Italy

Marine Malfroy
INSERM, EFS BFC, UMR 1098, RIGHT, Greffon-Hôte-Tumeur Interactions/Ingénierie Cellulaire et Génique, Université de Franche-Comté, Besançon, France

Joaquín Marco-Brualla
University of Zaragoza/Aragón Health Research Institute, Biochemistry and Molecular and Cell Biology, Zaragoza, Spain

Amélie Marguier
INSERM, EFS BFC, UMR 1098, RIGHT, Greffon-Hôte-Tumeur Interactions/Ingénierie Cellulaire et Génique, Université de Franche-Comté, Besançon, France

Isabel Marzo
Apoptosis, Immunity & Cancer Group, IIS Aragón, University of Zaragoza, Zaragoza, Spain

Ignacio Melero
Program of Immunology and Immunotherapy, Cima Universidad de Navarra; Navarra Institute for Health Research (IDISNA); Department of Oncology and Department of Immunology and Immunotherapy, Clínica Universidad de Navarra, Pamplona; Centro de Investigación Biomédica en Red de Cáncer (CIBERONC), Madrid, Spain; Nuffield Department of Medicine and Oxford Center for Immuno-Oncology, University of Oxford, Oxford, United Kingdom

Céline Mirjolet
Department of Radiation Oncology, Centre Georges-François Leclerc; Preclinical Radiation Therapy and Radiobiology Unit, Centre Georges-François Leclerc; TIReCS Team, UMR INSERM 1231, Dijon, France

Raquel Moreno-Loshuertos
University of Zaragoza/Aragón Health Research Institute, Biochemistry and Molecular and Cell Biology, Zaragoza, Spain

Véronique Morgand
Preclinical Radiation Therapy and Radiobiology Unit, Centre Georges-François Leclerc, Dijon, France

Etienne Moussay
Tumor Stroma Interactions Group, Department of Cancer Research, Luxembourg Institute of Health, Luxembourg, Luxembourg

Javier Naval
Apoptosis, Immunity & Cancer Group, IIS Aragón, University of Zaragoza, Zaragoza, Spain

Jerome Paggetti
Tumor Stroma Interactions Group, Department of Cancer Research, Luxembourg Institute of Health, Luxembourg, Luxembourg

Lucie Pavelková
Sotio; Department of Otorhinolaryngology and Head and Neck Surgery, First Medical Faculty, Motol University Hospital, Prague, Czech Republic

Klára Plačková
Sotio; Department of Otorhinolaryngology and Head and Neck Surgery, First Medical Faculty, Motol University Hospital, Prague, Czech Republic

Cristina Ripollés-Yuba
University of Zaragoza/Aragón Health Research Institute, Biochemistry and Molecular and Cell Biology, Zaragoza, Spain

Joan Salvador Russo-Cabrera
Program of Immunology and Immunotherapy, Cima Universidad de Navarra; Navarra Institute for Health Research (IDISNA), Pamplona, Spain

Víctor Sandá
Immunopathology Group, Biobizkaia Health Research Institute, Barakaldo, Spain

Ai Sato
Department of Radiation Oncology, Weill Cornell Medical College, New York, NY, United States

Alfonso Serrano-Del Valle
Apoptosis, Immunity & Cancer Group, IIS Aragón, University of Zaragoza, Zaragoza, Spain

Ruth Soler-Agesta
Department of Radiation Oncology, Weill Cornell Medical College, New York, NY, United States; University of Zaragoza/Aragón Health Research Institute, Biochemistry and Molecular and Cell Biology; Apoptosis, Immunity & Cancer Group, IIS Aragón, University of Zaragoza, Zaragoza, Spain

Mario Stan-Fontoba
Immunopathology Group, Biobizkaia Health Research Institute, Barakaldo, Spain

Alberto Susana
IRCCS Humanitas Research Hospital, Rozzano, Milan, Italy

Linn A. Syding
Sotio, Prague, Czech Republic

Francesca Urbani
Department of Oncology and Molecular Medicine, Istituto Superiore di Sanità (ISS), Rome, Italy

Samuel Valable
Université de Caen Normandie, CNRS, Normandie Université, ISTCT UMR6030, GIP Cyceron, Caen, France

Yaoyao Xie
INSERM, EFS BFC, UMR 1098, RIGHT, Greffon-Hôte-Tumeur Interactions/Ingénierie Cellulaire et Génique, Université de Franche-Comté, Besançon, France

Liwei Zhao
Equipe Labellisée Par La Ligue Contre Le Cancer, Université de Paris, Sorbonne Université, Centre de Recherche des Cordeliers, Institut Universitaire de France, Paris; Metabolomics and Cell Biology Platforms, Gustave Roussy, Université Paris Saclay, Villejuif, France

CHAPTER ONE

Assessment of translation rate in leukemic cells and immune cells of the microenvironment by OPP protein synthesis assay

Vanessa Klapp[a,b,†], Ozgu Gumustekin[a,†], Jerome Paggetti[a,‡], Etienne Moussay[a,‡], and Anne Largeot[a,‡,*]

[a]Tumor Stroma Interactions Group, Department of Cancer Research, Luxembourg Institute of Health, Luxembourg, Luxembourg

[b]Faculty of Science, Technology and Medicine, University of Luxembourg, Esch-sur-Alzette, Luxembourg

*Corresponding author: e-mail address: anne.largeot@lih.lu

Contents

† Contributed equally.
‡ Shared senior authorship.

Methods in Cell Biology, Volume 189
ISSN 0091-679X
https://doi.org/10.1016/bs.mcb.2024.06.006

Abstract

Despite being tightly regulated, messenger RNA (mRNA) translation, a manner in which cells control expression of genes and rapidly respond to stimuli, is highly dysfunctional and plastic in pathologies including cancer. Conversely, the investigation of molecular mechanisms whereby mRNA translation becomes aberrant in cancer, as well as inhibition thereof, become critical in developing novel therapeutic approaches. More specifically, in malignancies such as chronic lymphocytic leukemia in which aberrant global and transcript specific translation has been linked with poorer patient outcomes, targeting translation is a relevant approach, with various translation inhibitors under development. Here we elaborate on a protein synthesis assay by flow cytometry, O-propargyl-puromycin, demonstrating global mRNA translation rate with a variety of different applications including cell lines, primary cells or co-culture systems in vitro. This method provides a comprehensive tool in quantifying the rate of global mRNA translation in cancer cells, as well as that of the tumor microenvironment cells, or in response to inhibitory therapeutic agents while offering the possibility to simultaneously assess other cellular markers.

1. Introduction

Messenger RNA (mRNA) translation is a tightly regulated cellular process involved in protein synthesis and gene expression regulation, which plays a pivotal role in maintaining cellular homeostasis. In brief, mRNA translation can be divided into three distinct stages: initiation, elongation and termination (Fabbri et al., 2021). Translational control is fundamental for contributing to normal cellular functions, and thus is managed by different mechanisms installed along the line of the mRNA translation process, whereas the majority is occurring at the translation initiation step (Gebauer & Hentze, 2004; Jackson et al., 2010; Roux & Topisirovic, 2012). Conversely, mRNA translation alterations can occur at different stages and are often associated with several human pathologies including oncogenesis and cancer progression (Bjornsti & Houghton, 2004; Kochavi et al., 2023; Robichaud et al., 2019). In addition, aberrant mRNA translation was described to be a driver for cancer plasticity (Lee et al., 2021) and metastasis (Micalizzi et al., 2021).

Besides solid tumors, changes in mRNA translation are a common feature in hematological malignancies and lymphomas (Horvilleur et al., 2015; Topisirovic et al., 2003). In chronic lymphocytic leukemia (CLL) the mRNA translation machinery is affected at multiple different levels, which is ultimately linked to a more aggressive disease outcome or poorer prognosis (Klapp et al., 2023; Largeot et al., 2023). Specifically, members of the translation initiation complex, eukaryotic translation initiation factor 4F (eIF4F), typically involved in mRNA translation control, are often dysregulated in different types of cancers (Wu & Wagner, 2021) and thus constitute a promising therapeutic target (Bhat et al., 2015; Kovalski et al., 2022). In line with this, CLL cells present increased protein levels of eukaryotic translation initiation factor 4A (eIF4A), 4G (eIF4G) and phosphorylated 4E (eIF4E), which are implicated in the translation initiation machinery as members of the eIF4F complex (Largeot et al., 2023). Consistently, CLL cells from patients or from the spleen of Eμ-TCL1 mice (a transgenic mouse model for CLL) showed a higher quantification of mRNA translation at baseline compared to B cells derived from healthy donors or wild-type C56BL/6 mice, respectively (Largeot et al., 2023).

Apart from the aforementioned cancer cell specific aspects, dysregulation of mRNA translation can also affect the tumor microenvironment (TME) at various levels (Bartish et al., 2023). From an immunomodulatory perspective, mRNA translation was associated with immune checkpoint expression in cancer cells and thus provides rationale for potential combinatory treatment approaches, including mRNA translation inhibition and immune checkpoint blockade (Cerezo et al., 2021). Specifically, CD274 (best known as PD-L1) expression in tumor cells was shown to be dependent on the activation of the eIF4F complex upon interferon gamma (IFNγ) stimulation (Cerezo et al., 2018). Intriguingly, pharmacological inhibition or genetic silencing of members of the eIF4F complex lead to a decreased PD-L1 expression, resulting in an increased immune cell infiltration and reduced melanoma growth mediated by $CD8^+$ T cells (Cerezo et al., 2018). Similar effects were demonstrated in Epstein-Barr Virus-driven lymphoproliferative disease, whereas eIF4A inhibition by silvestrol, effectively blocked tumor progression while preserving anti-tumor immunity and cytotoxic activity of $CD8^+$ T cells (Patton et al., 2015). In models of aggressive liver cancer, inhibition of phosphorylation of eIF4E by a pharmacological inhibitor eFT508 was linked to down-regulation of PD-L1 following enhanced tumor control and prevention of metastasis formation (Xu et al., 2019). In accordance, absence of eIF4E phosphorylation in cells of the TME was associated with a loss of viability of prometastatic neutrophils, thus rendering genetically engineered tumor-bearing mice resistant to metastasis (Robichaud et al., 2018).

Inhibition of the translation initiation machinery can also be achieved by using the synthetic flavagline FL3, a known inhibitor of mRNA translation targeting prohibitin (PHB). FL3 treatment in preclinical models of CLL was also associated with reduced leukemic burden and extended overall survival in C56BL/6 mice adoptively transferred with Eμ-TCL1 splenocytes, without affecting the healthy, non-leukemic B cell population in vivo (Largeot et al., 2023). Interestingly, other PHB targeting compounds, such as fluorizoline was associated with a decrease in mRNA translation rate at sublethal doses (Largeot et al., 2023) however failed to inhibit CLL development in vivo (Wierz et al., 2018). CLL cell survival and proliferation are highly dependent on interactions with the immunosuppressive TME (Gargiulo et al., 2023; Pagano et al., 2023). Combining FL3 administration with immune checkpoint inhibition targeting programmed cell death 1 (PDCD1, best known as PD-1), thus further promoted increased anti-leukemic effects (Largeot et al., 2023). Conversely, translational control and reprogramming in immune cells can be essential for their activation and function, hence potential inhibition of their antitumor effect should be taken into consideration and evaluated for each individual setting when targeting mRNA translation as a therapeutic strategy in cancer (Bartish et al., 2023).

Inhibition of mRNA translation constitutes a powerful tool that showed efficient anticancer and immunomodulatory properties in various settings, including hematological malignancies. With the emerging importance of mRNA translation blockade and its implications on antitumor immunity alone or in combination with immunotherapy, assessing mRNA translation in tumor cells and cells from the TME constitutes an important tool. Overall, the translational status and mRNA that are actively engaged in translation can be evaluated by polysome and/or ribosome profiling (Cammas et al., 2022; Chassé et al., 2017; Jin & Xiao, 2018). In addition, the level of the interaction between the different members of the translation initiation machinery is a good surrogate of the cellular translation rate. An increased interaction is associated with elevated translation rate, whereas disruption of the translation initiation complex is representative for repressed mRNA translation. These interactions can be measured by proximity ligation assays (PLA) between eIF4G and eIF4E (Largeot et al., 2023;

Richter & Sonenberg, 2005). As an alternative to PLA, interplay between different key players of the translation initiation machinery can be assessed by NanoBRET, immunoprecipitation experiments and cap pull-down assay (Largeot et al., 2023). Nevertheless, these approaches come with limitations, including limited quantitative capabilities, the necessity for a high number of cells, or the need to sort cells to obtain information on heterogeneous populations.

Quantification of mRNA translation via O-propargyl-puromycin (OPP) protein synthesis assay by flow cytometry presents a simple and robust approach to determine the global mRNA translation and is applicable for cell lines, primary cultures or complex compositions containing a variety of different cells of the TME. This cytofluorometric approach offers readout-specific adaptations combining the OPP assay with different flow cytometry markers of choice, which makes this a versatile method enabling a deeper characterization of different key players in the TME. The technique harnesses a copper catalyzed click chemistry reaction linking an alkyne group of OPP to a fluorescently labeled azide (Liu et al., 2012). OPP is incorporated in the elongating peptide chain during protein synthesis as it mimics the 3′ end of a tRNA (Fig. 1A). As opposed to alternative (potentially hazardous) techniques, such as incorporation of radioactive labeled ^{35}S-methionine (Esposito & Kinzy, 2014) it offers an advantage in terms of safety and handling. Moreover, OPP can directly be added to the complete media without requiring specialized culture media as compared to labeled amino acid analogs such as homopropargylglycine (HPG) (Yousefi et al., 2021).

The herein presented chapter offers a detailed instruction for the investigation of the mRNA translation rate in human peripheral blood mononuclear cells (PBMCs) from CLL patients. This method provides a comprehensive analysis of mRNA translation in different subsets of leukemic and immune cells from primary PBMC of CLL patients, upon activation with TLR9-agonist Class B CpG ODN-2006 and treatment with the synthetic flavagline FL3. The protocol can be adapted and applied to determine mRNA translation in other cell lines, primary cells or co-culture systems in vitro. In addition, the protocol can also be employed to assess mRNA translation in lymphoid organs recovered from in vivo model after euthanasia, as described for the adoptive transfer Eu-TCL1 mouse model of CLL (Largeot et al., 2023).

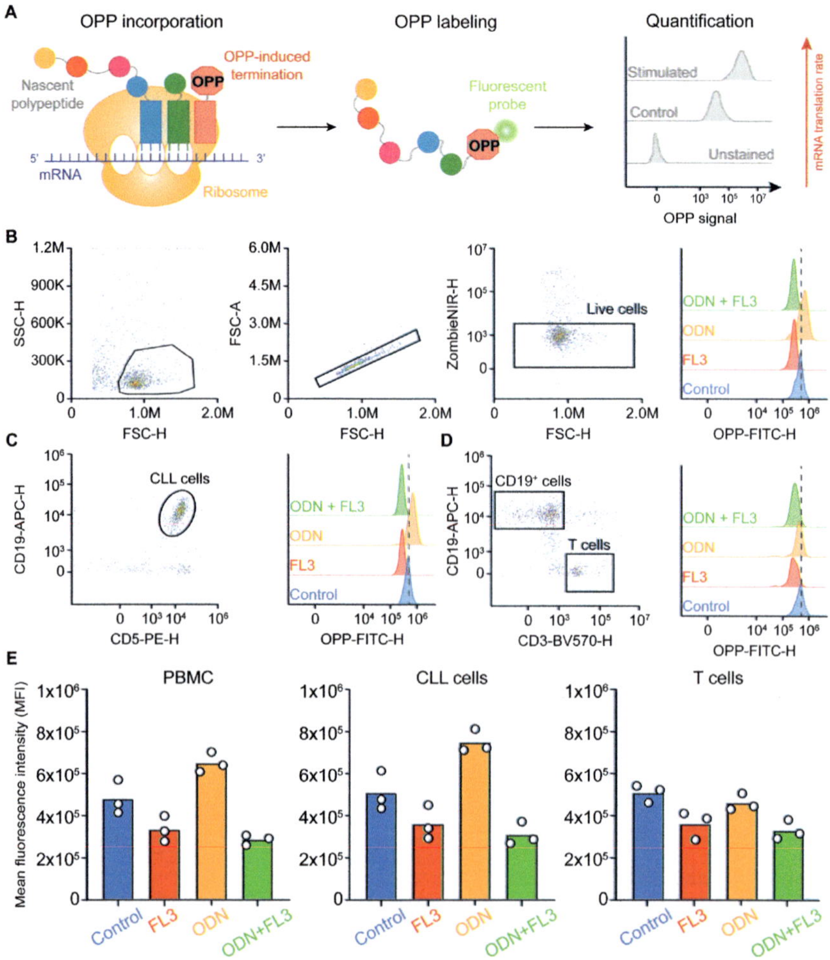

Fig. 1 O-Propargyl-puromycin (OPP) protein synthesis assay for human PBMCs from CLL patients. (A) OPP incorporation into newly synthesized peptides induces mRNA translation termination. Incorporated OPP can be labeled fluorescently and mRNA translation rate is assessed by flow cytometry. Intensity of the OPP fluorescent signal directly correlates with the mRNA translation rate. (B) Representative dot plots and the applied gating strategy of the flow cytometry analysis for human PBMCs from CLL patients are indicated. The percentage of CLL cells (C) or T cells (D) from live PBMCs are displayed and mRNA translation rate is presented in the depicted histograms. (E) Results of the mean fluorescence intensity (MFI) are reported as mean and individual data points. *Abbreviations*: OPP, O-propargyl-puromycin; CLL, chronic lymphocytic leukemia; PBMC, peripheral blood mononuclear cell; ODN, CpG ODN 2006.

2. Materials

2.1 Common disposables

- 15 mL conical-bottom centrifuge tubes (#188271, Greiner) (*see* **Note 1**)
- 50 mL conical-bottom centrifuge tubes (#227261, Greiner) (*see* **Note 1**)
- Leucosep tubes (#227290, Greiner) (*see* **Notes 1 and 2**)
- Malassez counting chamber (#0640610, Marienfeld Superior) (*see* **Notes 1 and 3**)
- Cryotubes (#5000-1020, Thermo Fisher) (*see* **Note 1**)
- T25 flasks (#690160, Greiner) (*see* **Note 1**)
- 96-Well plates, U-bottom (#650180, Greiner) (*see* **Note 1**)
- FACS tubes (#14-959-1A, Fisher Scientific) (*see* **Note 1**)

2.2 Cells and reagents

- PBMCs from CLL patients (*see* **Note 4**)
- Lymphocyte separation media (#392-0432, VWR) (*see* **Notes 5 and 6**)
- Trypan Blue Stain 0.4% (#T8154-100ML, Sigma Aldrich) (*see* **Notes 1 and 7**)
- Complete culture medium:
  - RPMI 1640 Medium, GlutaMAX™ Supplement (#61870036, Thermo Fisher Scientific) (*see* **Notes 1 and 5**)
  - 10% (v/v) fetal bovine serum (FBS) (#A5256801, Thermo Fisher Scientific) (*see* **Notes 1 and 5**)
  - 100 U/mL penicillin, 10,000 µg/mL streptomycin, 25 µg/mL of Amphotericin B (#HYCLSV30079.01, VWR) (*see* **Notes 1 and 8**)
  - 5 mM 2-[4-(2-hydroxyethyl) piperazin-1-yl] ethane sulfonic acid (HEPES; #15630-080, Life Technologies) (*see* **Notes 1 and 5**)
  - 50 µM 2-mercaptoethanol (#21985023, Thermo Fisher Scientific) (*see* **Notes 1 and 9**)
- CpG ODN 2006 (#130-100-105, Miltenyi Biotec) (*see* **Notes 1 and 5**)
- Synthetic flavagline FL3 (Qureshi et al., 2015) (*see* **Note 10**)
- Dimethyl sulfoxide (DMSO), 100 mL (#A994.1, Carl Roth GmbH) (*see* **Notes 1 and 9**)
- PBS, no calcium, no magnesium, pH 7.4 (1×) (#14190094, Life Technologies) (*see* **Notes 1 and 5**)
- *FACS buffer or* MACS buffer (#130-091-221, Miltenyi Biotec) (*see* **Notes 1 and 12**)

- FACS buffer: 10% FBS in PBS (*see* **Notes 1 and 5**)
- Antibodies APC anti-human CD19 (#302212, Biolegend), PE anti-human CD5 (#300608, Biolegend), BV570 anti-human CD3 (#300436, Biolegend), Zombie-NIR (#423106, Biolegend) (*see* **Notes 1 and 5**)
- OPP (#1416561-90-4, Bioconnect) (*see* **Notes 1 and 5**)
- Click-iT™ Plus OPP Alexa Fluor™ 488 Protein Synthesis Assay Kit (#C10456, Invitrogen)
  - Alexa Fluor® picolyl azide (*see* **Note 13**)
  - Click-iT® OPP Reaction Buffer (*see* **Note 14**)
  - Copper Protectant (*see* **Note 15**)
  - Click-iT® Reaction Buffer Additive (*see* **Note 5**)
- 4% (v/v) formaldehyde in PBS (#28908 Thermo Fisher Scientific) (*see* **Notes 1 and 16**)
- 0.5% (v/v) Triton X-100 in PBS (#X100, Sigma Aldrich) (*see* **Notes 1 and 17**)
- Distilled water (#T143.5 Carl Roth GmbH) (*see* **Notes 1 and 18**)

2.3 Equipment

- Laboratory biosafety cabinet (Class II), such as ESCO Airstream® Gen 3 Class II BSC- E Series (*see* **Note 1**)
- Humidified cell culture incubator at +37 °C with 5% CO_2, such as ESCO CelMate® CO_2 Incubator (*see* **Note 1**)
- Chemical fume hood, such as Possémé Sorbonne SPI VISIO (*see* **Note 1**)
- Standard bench top centrifuge, such as Eppendorf Centrifuge 5810/5810 R Benchtop Centrifuge (*see* **Note 1**)
- Flow cytometer equipped with lasers for at least 6 channel acquisition, such as Agilent NovoCyte Quanteon (*see* **Note 1**)

2.4 Software

- FlowJo, v. 10.6 (FlowJo LLC) (*see* **Note 19**)

3. Methods

3.1 Human PBMC isolation (*see* Note 20)

1. PBMCs are isolated from heparinized whole blood from CLL patients (*see* **Note 21**)

2. 50-mL Leucosep tubes are filled with 15 mL lymphocyte separation media and centrifuged at 1000 g for 30 s (*see* **Note 22**)
3. Whole blood is diluted 1:2 with PBS without Ca^{2+}/Mg^{2+} and added onto the porous membrane of the Leucosep tubes (*see* **Notes 23 and 24**)
4. Tubes are centrifuged at 1000*g* for 10 min at room temperature in a swinging bucket rotor with the break and acceleration of the centrifuge switched off (*see* **Note 25**)
5. After centrifugation the enriched PBMC containing fraction is separated in a thin layer located below the plasma and above the porous barrier (*see* **Note 26**)
6. Supernatant is carefully poured in a new 50 mL tube (*see* **Note 27**)
7. Tube containing the PBMC fraction is filled up with PBS without Ca^{2+}/Mg^{2+} and centrifuged at 350*g* for 5 min (*see* **Note 28**)
8. Do not discard supernatant. Supernatant is collected in a new 50 mL tube and washing step is repeated by centrifuging at 350*g* for 5 min to ensure maximal recovery of the cells
9. Supernatant is discarded and pellet is resuspended in complete media (*see* **Note 29**)
10. PBMCs are counted with trypan blue staining by any preferred method ideally adapted for PBMCs (*see* **Note 30**)
11. Cells can be used directly for any downstream experiments or applications (*see* Section 3.3), alternatively cells are cryopreserved as follows (*see* **Note 31**)
12. For cryopreservation, per cryovial a maximum of 100 million PBMCs are centrifuged at 350*g* for 5 min and resuspended in 750 μL of complete media. 750 μL of freezing media containing FBS + 20% DMSO is added drop by drop. Cell suspension is transferred to cryotubes and frozen via any preferred method. Cells are transferred to liquid nitrogen for long-term storage (*see* **Note 32**)
13. For thawing after cryopreservation, prewarmed complete media is added to the cryovial while gently resuspending and transferring the thawing solution to 9 mL of complete media. To remove the excess of DMSO, cells are centrifuged at 350*g* for 5 min (*see* **Note 33**)
14. After thawing, PBMCs are kept in the incubator for 1–2 h before assessing the viability and counting the cells for downstream experiments (*see* **Note 34**)

3.2 Preparation of treatment and staining stock solution (*see* Note 35)

1. FL3 is dissolved in DMSO to a 10 mM stock solution (*see* **Note 36**)
2. Intermediate working dilutions are prepared, diluting the 10 mM FL3 stock solution in a determined volume of DMSO prior to use (*see* **Note 37**)
3. OPP is dissolved in DMSO to a 20 mM stock solution (*see* **Note 38**)
4. 10× OPP Reaction Buffer Additive solution is prepared by adding 2 mL of ddH_2O while mixing thoroughly until the product is completely dissolved (*see* **Note 39**)

3.3 Cell culture and treatments (*see* Note 35)

1. Human PBMCs from CLL patients are used fresh or after thawing from cryopreservation (*see* Section 3.2)
2. PBMCs are seeded at a concentration of 4×10^6 cells/mL in 100 µL of complete media in a U-bottom 96-well plate (*see* **Note 40**)
3. For activation of CLL cells, CpG ODN 2006 is added at a final concentration of 7.5 µg/mL (*see* **Note 41**)
4. For mRNA translation inhibition treatment, FL3 is added at a final concentration of 100 nM (*see* **Note 42**)
5. Cells are activated with CpG ODN 2006 and/or treated with FL3 overnight (*see* **Note 43**)

3.4 OPP incorporation

1. After overnight treatment, OPP is added to each well at a final concentration of 20 µM (*see* **Note 44**)
2. Cells are incubated for OPP incorporation for 30 min at 37 °C (*see* **Note 45**)

3.5 Zombie NIR live/dead staining

1. After OPP incubation, cells in the 96-well plate are centrifuged at 400*g* for 5 min at 4 °C and the supernatant is removed (*see* **Note 46**)
2. Cells are washed with 1× PBS and centrifuged at 400*g* for 5 min at 4 °C
3. Viability staining mix is prepared by adding 0.2 µL of Zombie NIR fixable viability dye in 100 µL PBS per condition (*see* **Note 47**)
4. Cells are resuspended in 100 µL of the Zombie NIR staining mix and incubated for 15 min at room temperature protected from light (*see* **Note 48**)

3.6 Antibody staining

1. After Zombie NIR incubation, cells are centrifuged at 400*g* for 5 min at 4 °C
2. Supernatant is discarded and cells are washed with 100 μL PBS per well and centrifuged at 400*g* for 5 min at 4 °C (*see* **Note 46**)
3. Staining mix is prepared adding previously titrated amounts of CD19 and CD5 antibodies for CLL-defining markers as well as CD3 antibody for an additional T cell-defining marker in staining buffer (PBS + 10% FBS) (*see* **Note 49**)
4. Cells are resuspended in 100 μL of the prepared antibody staining mix and incubated for 30 min at 4 °C protected from light (*see* **Note 50**)

3.7 OPP staining (*see* Note 51)

1. After extracellular staining incubation, cells are centrifuged at 400*g* for 5 min
2. Supernatant is discarded and cells are washed with 100 μL PBS per well and centrifuged at 400*g* for 5 min (*see* **Note 46**)
3. Cells are resuspended in 100 μL of 4% formaldehyde for 15 min at room temperature protected from light for fixation (*see* **Note 52**)
4. After fixation, cells are centrifuged at 800*g* for 5 min (*see* **Note 53**)
5. Supernatant is discarded and cells are washed with PBS and centrifuged at 800*g* for 5 min (*see* **Note 46**)
6. Cells are resuspended in 100 μL of 0.5% (v/v) Triton X-100 in PBS and incubated for 15 min at room temperature protected from light (*see* **Note 52**)
7. During the incubation, OPP reaction cocktail is prepared freshly before use by combining 88 μL of 1× OPP Reaction Buffer, 2 μL of Copper Protectant, 0.25 μL of Alexa Fluor picolyl azide and 10 μL of 1× Reaction Buffer Additive (*see* **Notes 54–56**)
8. After permeabilization cells are centrifuged at 800 g for 5 min
9. Supernatant is discarded and cells are washed with PBS and centrifuged at 800*g* for 5 min (*see* **Note 46**)
10. Cells are resuspended in 25 μL of OPP reaction cocktail and incubated for 30 min at room temperature protected from light (*see* **Note 52**)
11. After OPP staining, cells are centrifuged at 800*g* for 5 min
12. Supernatant is discarded and cells are washed with PBS and centrifuged at 800*g* for 5 min (*see* **Note 46**)
13. If required, intracellular staining can be performed at this step

14. Cells are resuspended in 100 µL PBS per well and acquired on a flow cytometer (*see* **Note 57**)

3.8 Sample acquisition by flow cytometry

1. Multispectral flow cytometry is performed on a NovoCyte Quanteon (Agilent) equipped with four lasers (405 nm, 488 nm, 561 nm and 633 nm)
2. Data is acquired allowing a 5 channel acquisition used for PE (Y586/20), APC (R660/20), BV570 (V586/20), Alexa Fluor 488 (B530/30) and Zombie NIR (R780/60) besides FSC and SSC channels
3. Laser voltage and compensation are determined based on the acquisition of unstained and monocolor samples labeled with one fluorochrome only (*see* **Note 58**)

3.9 Data analysis

1. .fcs files can be analyzed using any standard flow cytometry software, such as FlowJo, v. 10.6 (FlowJo LLC) (*see* **Note 19**)
2. Compensation matrix is established based on monocolor labeled samples and applied to the experimental samples
3. For the assessment of the OPP translation rate the following gating strategy is employed: (1) Using FSC-H vs. SSC-H cells are selected based on granularity and morphology, then (2) doublets are excluded using FSC-A vs. FSC-H; (3) global translation rate can be displayed on a Alexa Fluor 488 B530-H histogram (Fig. 1B); (4) % of CLL cells can be deducted from the bi-parametric dot plots anti-CD19-APC-H vs anti-CD5-PE-H (Fig. 1C); (5) % of T cells can be determined on an anti-CD3-BV570-H plot (Fig. 1D); (6) additionally the translation rate can be determined for each subpopulation on a Alexa Fluor 488 B530-H histogram (Fig. 1C, D)
4. OPP translation rate can be quantified by calculating the mean fluorescent intensity (MFI) (Fig. 1E)

4. Notes

1. Catalog numbers and manufacturers are indicated as a reference, equivalent products can be obtained from different providers at similar cost.
2. Compared to standard PBMC isolation protocols, using Leucosep tubes results in a rapid and efficient isolation of PBMCs while the

porous barrier effectively prevents erythrocyte contamination. As an alternative, other products such as SepMate PBMC Isolation Tubes (#85450, Stemcell Technologies) can be used.

3. Automated cell counters validated for PBMCs counting (e.g., Countess 3 FL Automated Cell Counter, #AMQAF2000, Thermo Fisher Scientific) or other manual cell counting chambers (e.g., Neubauer chamber) can be used as an alternative.
4. All experiments involving human samples are conducted following the guidelines on ethical regulations in accordance with the declaration of Helsinki, approved by the national research ethics committee (*Luxembourg Comité National d'Ethique et de Recherche*, Luxembourg). For information regarding the regulation of work involving human samples, please refer to the guidelines established by the relevant authorities in your country and at your institution.
5. Lymphocyte separation media, RPMI 1640, FBS, HEPES, PBS, CpG ODN 2006, antibodies for flow cytometry, OPP and Click-iT® Reaction Buffer Additive are considered Non-hazardous (NONH) or Hazards not otherwise classified (HNOC) hence appropriate certified personal protective equipment (PPE) is recommended upon manipulation.
6. Other density gradient medium for the isolation of PBMCs such as Ficoll-Paque (#17144003 Cytiva) or Lymphoprep (#07851, Stemcell Technologies) can be used as an alternative.
7. Trypan Blue Stain may cause cancer (H350) and hence should be manipulated wearing appropriate certified PPE.
8. Penicillin may cause allergy or asthma symptoms or breathing difficulties upon inhalation (H334) and may cause allergic skin reaction (H317). Streptomycin is considered harmful if swallowed (H302) and suspected of damaging the unborn child upon exposure (H361d). Amphotericin B causes skin irritation (H315), serious eye irritation (H319) and may cause respiratory irritation (H335). Hence wearing appropriate certified PPE upon manipulation is recommended.
9. 2-mercaptoethanol may cause allergic skin reaction (H317) and is harmful to aquatic life (H402), thus should be manipulated wearing appropriate certified PPE and appropriate hazardous waste disposal should be ensured.
10. Other flavaglines known for inhibiting mRNA translation such as rocaglamide A or silvestrol can be used as an alternative.

11. DMSO is a combustible liquid (H227) and should be kept away from potential sources of ignition including, but not limited to heat, hot surfaces, open flames and sparks, and thus should be stored in appropriate and dedicated cabinets for flammable substances. DMSO adsorbs through the skin and hence should be manipulated wearing appropriate certified PPE.
12. MACS buffer contains sodium azide, which is considered harmful to aquatic life with long lasting effects (H412) and thus appropriate hazardous waste disposal must be ensured.
13. Alexa Fluor® picolyl azide causes mild skin irritation (H316), eye irritation (H320) and may cause respiratory irritation (H335) and hence should be manipulated wearing appropriate certified PPE. Alexa Fluor® picolyl azide is considered a combustible liquid (H227) and hence should be kept away from potential sources of ignition including, but not limited to heat, hot surfaces, open flames and sparks. This it should be stored in appropriate and dedicated cabinets for flammable substances.
14. Click-iT® OPP Reaction Buffer causes mild skin irritation (H216) and thus should be handled wearing appropriate certified PPE.
15. Copper Protectant causes mild skin irritation (H216) and hence should be manipulated wearing appropriate certified PPE. Copper Protectant is considered harmful to aquatic life (H402) and harmful to aquatic life with long lasting effects (H412), thus appropriate hazardous waste disposal must be ensured.
16. Formaldehyde 16% is harmful and toxic if swallowed (H301 + H302), in contact with skin (H311 + H312) and if inhaled (H331 + H332), causes severe skin burns and eye damage (H314), causes skin irritation (H315), may cause an allergic skin reaction (H317) and respiratory irritation (H335), is suspected of causing cancer (H351) and thus should be manipulated in a chemical fume hood with appropriate certified PPE. Formaldehyde is considered a flammable liquid and vapor (H226) and should be kept away from potential sources of ignition including, but not limited to heat, hot surfaces, open flames and sparks, and thus should be stored in appropriate and dedicated cabinets for flammable substances.
17. Triton X-100 is harmful if swallowed (H302), causes skin irritation (H315), causes serious eye damage (H318) and hence should be manipulated wearing appropriate certified PPE. Triton X-100 is considered very toxic to aquatic life (H400) and very toxic to aquatic life with long

lasting effects (H410) and thus appropriate hazardous waste disposal must be ensured.

18. Deionized water can be used as an alternative to DNase- and RNase-free distilled water.

19. Software provider and version are noted as a reference, other applications adapted to flow cytometry data analysis can be obtained from different providers (e.g., NovoExpress Software for the NovoCyte Platform).

20. Human PBMC isolation from CLL patients must be performed under a Class II biosafety cabinet to avoid any potential fungal or bacterial contamination and ensure the sterility of the sample for subsequent experiments. BSL2 safety regulations in accordance with the policies of your institute must be respected during the isolation of PBMCs from whole blood, thus wearing appropriate certified PPE is recommended.

21. Heparin prevents coagulation of the whole blood.

22. Lymphocyte separation media should be warmed to room temperature and kept protected from light.

23. It is crucial to use PBS without Ca^{2+}/Mg^{2+} to prevent whole blood from clotting.

24. For 50 mL Leucosep tubes it is recommended to use between 15 and 30 mL of diluted whole blood to ensure ideal efficiency of the density centrifugation.

25. Alternatively, tubes can be centrifuged for 15 min at 800*g* with brakes switched off.

26. Erythrocytes and granulocytes are pelleted below the porous membrane and thus avoiding contamination of the separated PBMC fraction

27. To avoid contamination of the enriched PBMCs with platelets from the plasma, it is recommended to carefully collect and discard some of the plasma layer located above the PBMC fraction

28. It is recommended to distribute a maximum of 10 mL of the previously isolated PBMC layer per 50 mL tube and fill up with 40 mL of PBS without Ca^{2+}/Mg^{2+} to optimally ensure washing the cell fraction from remains of the separation media.

29. The obtained cell fractions from both washing steps are pooled and resuspended in complete media

30. It is advised to stain cells with trypan blue solution to determine the viability of the PBMC cell suspension.

31. If the cells are used for downstream experiments, they should be kept at 37 °C in a 5% CO_2 incubator until use.

32. Freezing more than 100 million PBMCs per cryovial could reduce the viability of the cells. We recommend freezing the cells at −80 °C using isopropanol-containing freezing containers such as Mr. Frosty for 24 h before transferring the cryovials to liquid nitrogen.
33. To ensure a minimum loss of viability after thawing, it is recommended to proceed rapidly.
34. Decline in viability can occur after thawing, hence cells should be counted and viability should be assessed after 1–2 h.
35. The following steps must be performed under a Class II biosafety cabinet to avoid any potential fungal or bacterial contamination and ensure the sterility of the sample for subsequent experiments.
36. FL3 stock solutions are kept at −20 °C for 12 months.
37. For in vitro treatments, it is recommended to keep the total amount of DMSO at a maximum of 0.1% (v/v) to prevent detrimental effects of solvent to the cells, which can potentially impact experimental outcome.
38. OPP stock solution is kept at 4 °C, or alternatively at −20 °C for 12 months.
39. 10× solution of the Reaction Buffer Additive is stored at −20 °C for up to 1 year.
40. Using U-bottom 96-well plates is recommended to ensure cell proximity compared to F-bottom 96-well plates.
41. TLR9-agonist Class B CpG ODN-2006 is employed to activate CLL cells in the PBMC suspension, as an alternative CLL cells can also be stimulated via the B cell receptor (BCR) signaling pathway by coating Goat F(ab′)2 Anti-Human IgM (Southern Biotech) overnight at RT at a concentration of 50 μg/mL (Largeot et al., 2023).
42. For control conditions, an equivalent volume of DMSO without FL3 is added to the wells.
43. Overnight incubation at 37 °C in a humid 5% CO_2 incubator.
44. The 20 mM OPP stock solution is diluted 1:1000. When working in smaller volumes (e.g., 100 μL in this setting) it is advised to perform an intermediate dilution in complete media. Carefully move the plate to mix the OPP with the cell suspension.
45. 30 min incubation at 37 °C in a humid 5% CO_2 incubator. Longer incubation time might result in toxicity.
46. After pelleting the cells to the bottom of the 96-well U-bottom plate, the supernatant can be removed by quickly turning the plate upside down without detaching the cells.

47. The staining buffer should not contain Tris, BSA or FBS as this can interfere with the viability staining and might produce false positive signals
48. Staining time can be extended to up to 30 min.
49. It is recommended to titrate the antibodies to an optimized concentration for human PBMCs from CLL patients for extracellular staining. Other PBMC-defining markers can be included for a more detailed characterization of the present cell populations. Clones or fluorochromes are indicated as a reference, other alternatives can be used when compatible with the staining panel.
50. Staining time can be reduced to 20 min for extracellular staining with antibodies from the Biolegend brand, according to the manufacturer's instructions. However staining time should be adapted and adjusted to the experiment and antibodies used.
51. For further information, the user manual of the OPP staining kit, provided by Invitrogen can additionally be consulted.
52. Samples should be protected from light during incubations to minimize photobleaching.
53. Centrifugation speed can be increased up to 800*g* after fixation to ensure proper pelleting of the cells at the bottom of the plate.
54. 1× OPP Reaction Buffer can be prepared in advance by dilution the 10× stock solution 1:10 with distilled water. The 1× Reaction Buffer solution can be stored at 2–8 °C for up to 6 months.
55. 1× Reaction Buffer Additive solution must be prepared fresh from by diluting the 10× stock solution 1:10 with distilled water and used on the same day.
56. The OPP reaction cocktail must be used within 15 min of preparation.
57. If not acquiring immediately, cells must be kept at 4 °C protected from light for maximum of 24 h until acquisition, to preserve fluorescence and avoid doublet formation.
58. Including FMO (fluorescence minus one) samples, which include the whole panel except for one fluorochrome, is recommended to ensure proper gating control.

5. Concluding remarks

The protocol outlined here proposes a simple and versatile method in quantifying global mRNA translation rate in an array of different applications and sample types. The OPP assay presents a reliable method in assessing efficiency when acquiring a deeper understanding of the molecular mechanisms of action of therapeutic agents inhibiting mRNA translation via the translation initiation machinery. This assay exploits the incorporation of OPP, a puromycin analog, into the newly synthesized proteins and offers a direct indication of active mRNA translation in the cells. The nascently incorporated OPP is fluorescently labeled via a click chemistry reaction and detected by flow cytometry or microscopy. The main advantages of this method are high sensitivity and ease of use with minimal hands-on time (Li et al., 2022). Unlike traditional methods for studying protein synthesis, it is non-radioactive and is therefore safer and more convenient. Further, it is versatile in array of suitable samples and compatible with different downstream applications, including fluorescence microscopy, flow cytometry, and high-content screening (Li et al., 2022). In addition, OPP assay has been deemed efficient in vivo by direct administration to mice followed by bone marrow harvest (Hidalgo & Signer, 2019). Some limitations may include potential cellular toxicity, limited detection range, time dependence and background, due to nonspecific labeling or detection. Despite these limitations, the Click-iT® Plus OPP Protein Synthesis Assay remains a valuable tool for studying protein synthesis dynamics in cells, especially when used judiciously and in conjunction with complementary experimental approaches. We recommend carefully considering these limitations and optimizing assay conditions, to ensure accurate and meaningful interpretation of the results obtained. Nevertheless, alternative methods worth mentioning include the HPG Click-iT, utilizing HPG (L-homopropargylglycine), a methionine analog, which is incorporated into newly synthesized proteins during translation (Li et al., 2022; Morral et al., 2020). Importantly, this method requires changing of media prior to starting the assay. Similarly to the HPG, the Azidohomoalanine (AHA) or L-azidohomoalanine (L-AHA) is another methionine analog that is incorporated into newly synthesized proteins, quantifiable via fluorescent labeling (Ma & Yates, 2018). Further, the incorporation of radioactive sulfur-35-labeled methionine (^{35}S-methionine) can also be used to quantify nascent protein synthesis. However, this method involves the use of radioactive materials and ^{35}S-methionine has a relatively short half-life, which limits the duration of experiments and the ability to track protein synthesis over extended periods of time (Esposito & Kinzy, 2014). Other incorporation-based methods exist, that are alternatively directly detectable with mass cytometry, such as L-2-tellurienylalanine (TePhe), a nontoxic and stable noncanonical amino acid (Bassan et al., 2019), which can also be combined with other cell defining markers (Gonder et al., 2020). Another method making use of multi-epitope tags and antibody-based fluorescent probes, nascent chain tracking (NCT), has been described to real-time quantify protein synthesis at a single-molecule level (Morisaki et al., 2016). Finally, alternatives possible to employ in detecting rate of protein synthesis are; Proximity Ligation Assay (PLA) capable of detecting protein interactions, Polysome Profiling monitoring translational activity of mRNAs, Pulse-Chase Labeling with Radioactive Amino Acids and Stable Isotope Labeling with Amino Acids in Cell Culture (SILAC) (Largeot et al., 2023).

Acknowledgments

This work was supported by grants from the Luxembourg National Research Fund (FNR) and Fondation Cancer to VK, OG, EM, JP and AL (PRIDE19/14254520/i2TRON, C20/BM/14582635, C20/BM/14592342, and C23/BM/17987391).

Competing interests

Not applicable.

References

Bartish, M., et al. (2023). The role of eIF4F-driven mRNA translation in regulating the tumour microenvironment. *Nature Reviews. Cancer*, *23*, 408–425. https://doi.org/10.1038/s41568-023-00567-5.

Bassan, J., et al. (2019). TePhe, a tellurium-containing phenylalanine mimic, allows monitoring of protein synthesis in vivo with mass cytometry. *Proceedings of the National Academy of Sciences of the United States of America*, *116*, 8155–8160. https://doi.org/10.1073/pnas.1821151116.

Bhat, M., et al. (2015). Targeting the translation machinery in cancer. *Nature Reviews. Drug Discovery*, *14*, 261–278. https://doi.org/10.1038/nrd4505.

Bjornsti, M. A., & Houghton, P. J. (2004). Lost in translation: Dysregulation of cap-dependent translation and cancer. *Cancer Cell*, *5*, 519–523. https://doi.org/10.1016/j.ccr.2004.05.027.

Cammas, A., et al. (2022). Analysis of mRNA translation by polysome profiling. *Methods in Molecular Biology*, *2404*, 69–81. https://doi.org/10.1007/978-1-0716-1851-6_4.

Cerezo, M., et al. (2018). Translational control of tumor immune escape via the eIF4F-STAT1-PD-L1 axis in melanoma. *Nature Medicine*, *24*, 1877–1886. https://doi.org/10.1038/s41591-018-0217-1.

Cerezo, M., et al. (2021). The role of mRNA translational control in tumor immune escape and immunotherapy resistance. *Cancer Research*, *81*, 5596–5604. https://doi.org/10.1158/0008-5472.Can-21-1466.

Chassé, H., et al. (2017). Analysis of translation using polysome profiling. *Nucleic Acids Research*, *45*, e15. https://doi.org/10.1093/nar/gkw907.

Esposito, A. M., & Kinzy, T. G. (2014). In vivo [35S]-methionine incorporation. *Methods in Enzymology*, *536*, 55–64. https://doi.org/10.1016/b978-0-12-420070-8.00005-2.

Fabbri, L., et al. (2021). The plasticity of mRNA translation during cancer progression and therapy resistance. *Nature Reviews. Cancer*, *21*, 558–577. https://doi.org/10.1038/s41568-021-00380-y.

Gargiulo, E., et al. (2023). Extracellular vesicle secretion by leukemia cells in vivo promotes CLL progression by hampering antitumor T-cell responses. *Blood Cancer Discovery*, *4*, 54–77. https://doi.org/10.1158/2643-3230.Bcd-22-0029.

Gebauer, F., & Hentze, M. W. (2004). Molecular mechanisms of translational control. *Nature Reviews. Molecular Cell Biology*, *5*, 827–835. https://doi.org/10.1038/nrm1488.

Gonder, S., et al. (2020). Method for the analysis of the tumor microenvironment by mass cytometry: Application to chronic lymphocytic leukemia. *Frontiers in Immunology*, *11*, 578176. https://doi.org/10.3389/fimmu.2020.578176.

Horvilleur, E., et al. (2015). Cap-independent translation in hematological malignancies. *Frontiers in Oncology*, *5*, 293. https://doi.org/10.3389/fonc.2015.00293.

Jackson, R. J., et al. (2010). The mechanism of eukaryotic translation initiation and principles of its regulation. *Nature Reviews. Molecular Cell Biology*, *11*, 113–127. https://doi.org/10.1038/nrm2838.

Jin, H. Y., & Xiao, C. (2018). An integrated polysome profiling and ribosome profiling method to investigate in vivo translatome. *Methods in Molecular Biology*, *1712*, 1–18. https://doi.org/10.1007/978-1-4939-7514-3_1.

Hidalgo San Jose, L., & Signer, R. A. J. (2019). Cell-type-specific quantification of protein synthesis in vivo. *Nature Protocols*, *14*, 441–460. https://doi.org/10.1038/s41596-018-0100-z.

Klapp, V., et al. (2023). Targeting mRNA translation aberrations: A novel approach for therapy in chronic lymphocytic leukemia. *Cancer Communications*, *43*, 1373–1376. https://doi.org/10.1002/cac2.12493.

Kochavi, A., et al. (2023). Proteome diversification by mRNA translation in cancer. *Molecular Cell*, *83*, 469–480. https://doi.org/10.1016/j.molcel.2022.11.014.

Kovalski, J. R., et al. (2022). Protein synthesis control in cancer: Selectivity and therapeutic targeting. *The EMBO Journal*, *41*, e109823. https://doi.org/10.15252/embj.2021109823.

Largeot, A., et al. (2023). Inhibition of MYC translation through targeting of the newly identified PHB-eIF4F complex as a therapeutic strategy in CLL. *Blood*, *141*, 3166–3183. https://doi.org/10.1182/blood.2022017839.

Lee, L. J., et al. (2021). Cancer plasticity: The role of mRNA translation. *Trends in Cancer*, *7*, 134–145. https://doi.org/10.1016/j.trecan.2020.09.005.

Li, Y., et al. (2022). Click-iT (®) Plus OPP Alexa Fluor (®) Protein Synthesis Assay in embryonic cells. *Bio-Protocol*, *12*, e4441. https://doi.org/10.21769/BioProtoc.4441.

Liu, J., et al. (2012). Imaging protein synthesis in cells and tissues with an alkyne analog of puromycin. *Proceedings of the National Academy of Sciences of the United States of America*, *109*, 413–418. https://doi.org/10.1073/pnas.1111561108.

Ma, Y., & Yates, J. R., 3rd. (2018). Proteomics and pulse azidohomoalanine labeling of newly synthesized proteins: What are the potential applications? *Expert Review of Proteomics*, *15*, 545–554. https://doi.org/10.1080/14789450.2018.1500902.

Micalizzi, D. S., et al. (2021). Translational regulation of cancer metastasis. *Cancer Research*, *81*, 517–524. https://doi.org/10.1158/0008-5472.Can-20-2720.

Morisaki, T., et al. (2016). Real-time quantification of single RNA translation dynamics in living cells. *Science*, *352*, 1425–1429. https://doi.org/10.1126/science.aaf0899.

Morral, C., et al. (2020). Protocol for efficient protein synthesis detection by click chemistry in colorectal cancer patient-derived organoids grown in vitro. *STAR Protocols*, *1*, 100103. https://doi.org/10.1016/j.xpro.2020.100103.

Pagano, G., et al. (2023). Interleukin-27 potentiates CD8+ T-cell-mediated antitumor immunity in chronic lymphocytic leukemia. *Haematologica*, *108*, 3011–3024. https://doi.org/10.3324/haematol.2022.282474.

Patton, J. T., et al. (2015). The translation inhibitor silvestrol exhibits direct anti-tumor activity while preserving innate and adaptive immunity against EBV-driven lymphoproliferative disease. *Oncotarget*, *6*, 2693–2708. https://doi.org/10.18632/oncotarget.2098.

Qureshi, R., et al. (2015). FL3, a synthetic flavagline and ligand of prohibitins, protects cardiomyocytes via STAT3 from doxorubicin toxicity. *PLoS One*, *10*, e0141826. https://doi.org/10.1371/journal.pone.0141826.

Richter, J. D., & Sonenberg, N. (2005). Regulation of cap-dependent translation by eIF4E inhibitory proteins. *Nature*, *433*, 477–480. https://doi.org/10.1038/nature03205.

Robichaud, N., et al. (2018). Translational control in the tumor microenvironment promotes lung metastasis: Phosphorylation of eIF4E in neutrophils. *Proceedings of the National Academy of Sciences of the United States of America*, *115*, E2202–e2209. https://doi.org/10.1073/pnas.1717439115.

Robichaud, N., et al. (2019). Translational control in cancer. *Cold Spring Harbor Perspectives in Biology*, *11*. https://doi.org/10.1101/cshperspect.a032896.

Roux, P. P., & Topisirovic, I. (2012). Regulation of mRNA translation by signaling pathways. *Cold Spring Harbor Perspectives in Biology*, *4*. https://doi.org/10.1101/cshperspect.a012252.

Topisirovic, I., et al. (2003). Aberrant eukaryotic translation initiation factor 4E-dependent mRNA transport impedes hematopoietic differentiation and contributes to leukemogenesis. *Molecular and Cellular Biology*, *23*, 8992–9002. https://doi.org/10.1128/mcb.23.24.8992-9002.2003.

Wierz, M., et al. (2018). The prohibitin-binding compound fluorizoline induces apoptosis in chronic lymphocytic leukemia cells ex vivo but fails to prevent leukemia development in a murine model. *Haematologica*, *103*, e154–e157. https://doi.org/10.3324/haematol.2017.175349.

Wu, S., & Wagner, G. (2021). Deep computational analysis details dysregulation of eukaryotic translation initiation complex eIF4F in human cancers. *Cell Systems*, *12*, 907–923. e906. https://doi.org/10.1016/j.cels.2021.07.002.

Xu, Y., et al. (2019). Translation control of the immune checkpoint in cancer and its therapeutic targeting. *Nature Medicine*, *25*, 301–311. https://doi.org/10.1038/s41591-018-0321-2.

Yousefi, R., et al. (2021). Monitoring mitochondrial translation in living cells. *EMBO Reports*, *22*, e51635. https://doi.org/10.15252/embr.202051635.

CHAPTER TWO

Generation of transmitochondrial cybrids in cancer cells

Ruth Soler-Agesta[a,b], Cristina Ripollés-Yuba[a], Joaquín Marco-Brualla[a], Raquel Moreno-Loshuertos[a], Ai Sato[b], Manuel Beltrán-Visiedo[b], Lorenzo Galluzzi[b,c,d,*,†], and Alberto Anel[a,*,†]

[a]University of Zaragoza/Aragón Health Research Institute, Biochemistry and Molecular and Cell Biology, Zaragoza, Spain
[b]Department of Radiation Oncology, Weill Cornell Medical College, New York, NY, United States
[c]Sandra and Edward Meyer Cancer Center, New York, NY, United States
[d]Caryl and Israel Englander Institute for Precision Medicine, New York, NY, United States
*Corresponding authors: e-mail address: deadoc80@gmail.com; anel@unizar.es

Contents

Abstract

At odds with historical views suggesting that mitochondrial functions are largely dispensable for cancer cells, it is now clear that mitochondria have a major impact on malignant transformation, tumor progression and response to treatment.

† Co-senior authors.

Methods in Cell Biology, Volume 189
ISSN 0091-679X
https://doi.org/10.1016/bs.mcb.2024.05.010

Mitochondria are indeed critical for neoplastic cells not only as an abundant source of ATP and other metabolic intermediates, but also as gatekeepers of apoptotic cell death and inflammation. Interestingly, while mitochondrial components are mostly encoded by nuclear genes, mitochondria contain a small, circular genome that codes for a few mitochondrial proteins, ribosomal RNAs and transfer RNAs. Here, we describe a straightforward method to generate transmitochondrial cybrids, i.e., cancer cells depleted of their mitochondrial DNA and reconstituted with intact mitochondria from another cellular source. Once established, transmitochondrial cybrids can be stably propagated and are valuable to dissect the specific impact of the mitochondrial genome on cancer cell functions.

1. Introduction

More than a century ago, the German physiologist Otto Warburg was the first to report that malignant lesions exhibit a higher propensity to take up glucose as compared to normal tissues, a process that is commonly known as the "Warburg effect" (Koppenol et al., 2011; Thompson et al., 2023; Warburg et al., 1924; Warburg & Minami, 1923). Such a seminal discovery not only laid the foundation for an improved understanding of how cancer cells reconfigure their metabolism in support of tumor progression and resistant to therapy, de facto spearheading a vivid line of investigation attempting to target tumor metabolism for therapeutic purposes (Stine et al., 2022), but also provided the rationale for the development of a diagnostic tool that is commonly employed for clinical cancer management, [^{18}F]-fluorodeoxyglucose positron emission tomography ([^{18}F]-FDG PET) scans (Schwenck et al., 2023; Yankeelov et al., 2014). That said, Warburg postulated that malignant cells largely rely on glucose for bioenergetic metabolism as a consequence of mitochondrial dysfunction (Warburg, 1956a, 1956b), a possibility that has now been invalidated (Vaupel & Multhoff, 2021; Vyas et al., 2016; Zong et al., 2016). Indeed, accrued glucose uptake by malignant cells appears to mainly fuel anabolic (rather catabolic) metabolism in support of lipid and nucleotide biosynthesis (which are crucial for accelerated proliferation), mitochondrial ATP remaining the major source of energy (DeBerardinis & Chandel, 2020; Locasale et al., 2011; Lunt & Vander Heiden, 2011; Vander Heiden et al., 2009). Moreover, mitochondrial metabolism at large—including catabolic and anabolic components thereof—is fundamental for all steps of oncogenesis, including malignant transformation, tumor progression and response to treatment (Dora et al., 2023; Hanahan, 2022; Hanahan & Weinberg, 2011; Oldan et al., 2023; Petroni et al., 2021;

Porporato et al., 2018). Finally, mitochondrial integrity is paramount for the inhibition of cell-intrinsic (as mediated by regulated cell death) (Diepstraten et al., 2022; Vitale et al., 2023; Yuan & Ofengeim, 2023) as well as cell-extrinsic (as mediated by the host immune system) oncosuppression (Catalán et al., 2015; Kroemer et al., 2023; Marchi et al., 2023; Newman & Shadel, 2023; Sprooten et al., 2023; Vringer & Tait, 2023; Yamazaki et al., 2020a).

Of note, while the majority of mitochondrial constituents including all lipids and the vast majority of proteins is synthesized based on nuclear genetic information, mitochondria contain a small, circular genome that replicated independent of genomic DNA and encode a few mitochondrial proteins, ribosomal RNAs and transfer RNAs (Medini et al., 2020; Neupert, 2016; Rackham & Filipovska, 2022). Moreover, (1) mtDNA inheritance in daughter cells is largely stochastic (Gustafsson et al., 2016); and (2) most importantly, cell contains several mitochondrial DNA (mtDNA) molecules, which are not necessarily identical to each other (a scenario commonly referred to as heteroplasmy) (Stewart & Chinnery, 2021). Coupled to the fact that mtDNA replication is more prone to errors than genomic DNA (gDNA) replication (Roy et al., 2022), these observations imply that individual mtDNA molecules have a high propensity to accumulate mutations, which can stochastically accumulate in daughter cells with a potentially pathogenic decrease in heteroplasmy (Craven et al., 2017; Kopinski et al., 2021; Yang et al., 2019).

Importantly, while the CRISPR/Cas9 technology has rendered experimental gDNA alterations routinary not only in cultured cells but also in animals (Chavez et al., 2023; Madigan et al., 2023; Villiger et al., 2024; Zhang et al., 2021), artificially altering mtDNA in cells and organisms remains challenging. To some degree, such obstacle can be partially circumvented by the generation of "transmitochondrial cybrids," i.e., cells depleted of their own mtDNA (generally referred to as rho^0 or ρ^0 cells) that are reconstituted with intact mitochondria from another cellular source (Cavaliere et al., 2022; Soler-Agesta et al., 2023) (Fig. 1). Here, we detail a straightforward protocol to generate prostate cancer transmitochondrial cybrids based on human prostate cancer DU-145 cells as recipient cells and human prostate cancer 22Rv1 cells as donor cells. With minimal variations, we expect this method to be simply adaptable to other recipient and donor cancer cell lines of human and murine origin.

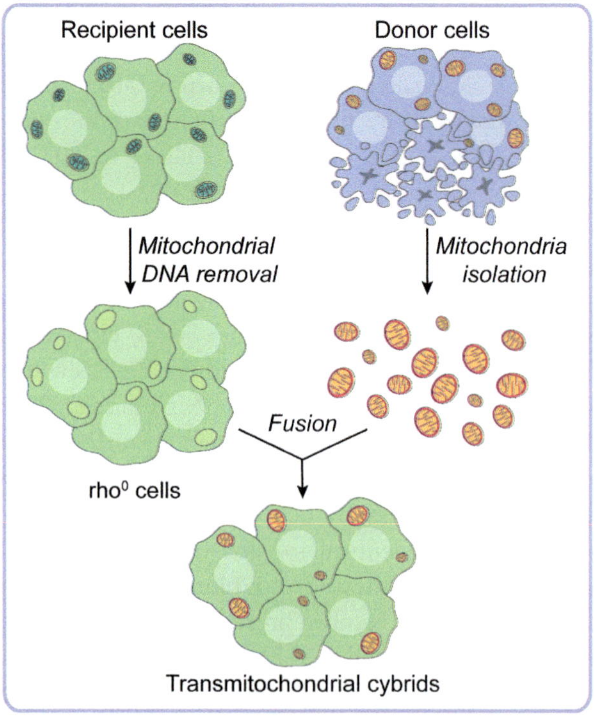

Fig. 1 Generation of transmitochondrial cybrids. Recipient cells are treated with sublethal concentrations of ethidium bromide to eliminate mitochondrial DNA (mtDNA) and generate rho^0 cells. Once rho^0 recipient cells are established, intact mitochondria are isolated from donor cells and used to reconstitute a functional mitochondrial network in transmitochondrial cybrids.

2. Materials

2.1 Common disposables (*see Note 1*)

- Cell culture flasks 25 cm^2 (#90026, Techno Plastic Products)
- Cell culture flasks 75 cm^2 (#90076, Techno Plastic Products)
- Corning® 96-well Clear Flat Bottom Polystyrene TC-treated Microplates, 5 per Bag, with Lid, Sterile (#3598, Corning)
- Costar® 24-well Clear TC-treated Multiple Well Plates, Individually Wrapped, Sterile (#09–761-146, Corning)
- Falcon® 15 mL High Clarity PP Centrifuge Tube, Conical Bottom, with Dome Seal Screw Cap, Sterile (#352096, Corning)
- Falcon® 50 mL High Clarity PP Centrifuge Tube, Conical Bottom, with Dome Seal Screw Cap, Sterile (#352070, Corning)
- Flex-Tube® 1.5 mL (#022364111, Eppendorf)

2.2 Cells and reagents (*see Note 1*)

- 10× TBE Buffer (Tris/Boric Acid/EDTA), 1 L (#1610733, BioRad)
- Buffer A: 10 mM Tris, 1 mM EDTA, 0.32 M sucrose, pH 7.4
- Complete DMEM: Gibco™ Dulbecco's Modified Eagle's Medium (DMEM), high glucose (4.5 g/L), GlutaMAX™ Supplement, supplemented with 10% FBS, 10,000 U/mL penicillin and 10 mg/mL streptomycin
- Complete RMPI 1640 medium: Gibco™ RPMI 1640 Medium, supplemented with 10% FBS, 10,000 U/mL penicillin and 10 mg/mL streptomycin.
- Ethanol 99% (#100983, Supelco)
- Ethidium Bromide (#E8751, Sigma-Aldrich) (***see Note 2***)
- Fetal Bovine Serum (FBS, #F7524, Sigma-Aldrich)
- Gibco™ Dulbecco's Modified Eagle's Medium (DMEM), high glucose (4.5 g/L), GlutaMAX™ Supplement (#10566016, Thermo-Fisher Scientific)
- Gibco™ RPMI 1640 Medium (#11875093, Thermo-Fisher Scientific)
- Human 22Rv1 prostate cancer cells (#CRL-2505, ATCC)
- Human DU-145 prostate cancer cells (#HTB-81, ATCC)
- LightCycler® FastStart DNA Master^PLUS SYBR Green I (#03515869001, Roche)
- MOPS (#M1254, Sigma-Aldrich)
- Nuclease free water (#MC1191, Promega)
- Penicillin-Streptomycin, 10,000 U/mL Penicillin, 10 mg/mL Streptomycin (#P06–07100, PAN-Biotech)
- Polyethylene glycol (PEG) solution (#P7181, Sigma-Aldrich)
- QIAamp® DSP DNA Blood Mini Kit (#61104, Qiagen)
- qPCR primers (Table 1)
- *SacI*-HF® restriction enzyme kit (#R3156S, New England BioLabs®)
- Sodium pyruvate (#P5280, Sigma-Aldrich)
- Sucrose (#S9378, Sigma-Aldrich)
- TopVision Agarose (#R0491, ThermoFisher Scientific)
- Trypan Blue Solution, 0.4% (#T10282, Invitrogen)
- Trypsin 0.25%/1 mM EDTA 4Na in HBSS, w/o: Ca and Mg, w: Phenol red (#P10-029100, PAN-Biotech)
- Uridine (#U3750, Sigma-Aldrich)

2.3 Equipment (*see Note 1*)

- 5331 Mastercycler Gradient Thermal Cycler (Eppendorf)
- Azure 600® Imager (Azure Biosystems)

Table 1 PCR primers.

Gene	Primer	Position	Sequence
MT-CO2 (NC_012920)	MT-CO2-FWD	7858–7877	CGATCCCTCCCTTACCATCA
	MT-CO2-REV	7926–7904	CCGTAGTCGGTGTACTCGTAGGT
SDHA (NC_000005.10)	SDHA-FWD	2641–2661	TCTCCAGTGGCCAACAGTGTT
	SDHA-REV	2712–2693	GCCCTCTTGTTCCCATCAAC

Abbreviations: FWD, forward; MT-CO2, mitochondrially encoded cytochrome *c* oxidase II; REV, reverse; SDHA, succinate dehydrogenase complex flavoprotein subunit A.

- CK inverted microscope (Olympus)
- Fisherbrand™ Horizontal Electrophoresis Systems (#FB-SB-710, Fisher Scientific)
- Homogenizer, motor-driven rotor (Bosch)
- LightCycler® 2.0 System (Roche)
- MCO-18AIC (UV) CO_2 Incubator for Cell Culture (Sanyo)
- MicrofugeR 22R Centrifuge (Beckman Coulter™)
- NanoVue™ Plus Spectrophotometer (GE Healthcare)
- Telstar™ PV-30/70 Vertical Laminar Air Flow Cabinet (Telstar™)
- Tissue Grinders-homogenizer, Glass Vessel, 2 mL (#196102, Deltalab)
- Tissue Grinders-homogenizer, Plunger-serrated Tip, 2 mL (#196302, Deltalab)

2.4 Software (*see Note 1*)

- Excel (Microsoft)
- LightCycler® v. 4.1 (Roche)
- Primer Express 2.0 (Applied Biosystems)
- Prism v. 10.0.1 (GraphPad)

3. Methods

3.1 rho^0 cell generation (*see Notes 3 and 4*)

1. Wild-type (WT) DU-145 cells are maintained in culture using complete DMEM supplemented with 50 μg/mL uridine and 110 μg/mL pyruvate (***see Notes 5 and 6***).
2. Treat 2–3 × 10^6 DU-145 cells with 500 ng/mL ethidium bromide for a total of 10 days, refreshing ethidium bromide-containing medium every 48 h (***see Notes 7 and 8***).

3.2 Quantification of mtDNA by qPCR

The degree of mtDNA depletion in DU-145 cells exposed to ethidium bromide is assessed by qPCR based on nuclear and mitochondrial gene primers (***see Notes 9–11***).

3.2.1 DNA isolation

1. Collect 1–5 × 10^6 DU-145 cells in a 15 mL tube and centrifuge the cell suspension at 520 × *g* for 5 min.
2. Discard the supernatant, resuspend the pellet in 5 mL PBS, centrifuge the cell suspension at 520 × *g* for 5 min.

3. Discard the supernatant and resuspend the pellet in 200 μL PBS.
4. Add 20 μL proteinase K (20 mg/mL) and 200 μL buffer AL as provided in the DNA Extraction Kit and vortex for 15 s.
5. Incubate at 56 °C for 10 min.
6. Add 200 μL ethanol (96–100%) and vortex for 15 s.
7. Transfer the content of the tube into a column as provided by the DNA Extraction Kit and centrifuge at 6000 × *g* for 1 min.
8. Transfer the column to a clean 2 mL collection tube as provided by the DNA Extraction Kit and discard the tube with the filtrate.
9. Add 500 μL buffer AW1 as provided by the DNA Extraction Kit, centrifuge at 6000 × *g* for 1 min and transfer the column to a clean tube.
10. Add 500 μL buffer AW2 as provided by the DNA Extraction Kit, centrifuge at maximum speed for 3 min and transfer the column to a clean tube.
11. Centrifuge at maximum speed for 3 min to remove any trace of ethanol.
12. Transfer the column to a 1.5 mL microcentrifuge tube, add 200 μL buffer AE as provided by the DNA Extraction Kit (***see Note 12***) and incubate for 1 min at room temperature (RT).
13. Recover DNA by centrifuging at 6000 × *g* for 1 min and store the sample at −20 °C (***see Note 13***).

3.2.2 mtDNA quantification

1. Quantify DNA concentration and purity with a NanoVue™ Plus Spectrophotometer.
2. Dilute samples to a concentration of 5 ng/μL using PCR-grade distilled water.
3. Prepare a master mix composed of 4 μL LightCycler® FastStart DNA Master^{PLUS} SYBR Green I, 0.5 μL primer mix (5 μM for each primer) and 12.5 μL PCR-grade distilled water, per reaction (***see Note 14***).
4. Distribute the mix in a precooled qPCR capillary tube, add 3 μL DNA template (5 ng/μL) up to 20 μL total rection volume and centrifuge at 1000 × *g* for 1 min to enable reagents mix with the template (***see Note 15***).
5. Insert the capillary tube in the LightCycler® 2.0 System and set thermal protocol (Table 2).
6. Results are obtained as cycle threshold (CT) for both genomic and mitochondrial amplicons. To evaluate mtDNA abundance, calculate the average of CT numbers obtained on genomic and mitochondrial amplicons, and plug them into the formula: $2^{(\text{average gDNA CT}-\text{ average mtDNA CT})}$. A cell line is considered rho^0 when CT obtained with mitochondrial primers is equivalent to CT in negative control conditions (Fig. 2).

Table 2 Thermal protocol for mtDNA quantification by qPCR.

Step	Temperature	Time	Cycles
Initial denaturation	95 °C	10 min	1 cycle
Quantification	95 °C	10 s	55 cycles
	60 °C[a]	10 s	
	72 °C	10 s	
Melting curve	95 °C	1 s	1 cycle
	65 °C	15 s	
	65 ⟶ 98 °C	330 s (rate: 0.1 °C/s)	
Cooling	40 °C	30 s	1 cycle

[a]This temperature depends on the primer sequence and length, and it can be calculated by using the following equation: Temperature = 2 (A + T) + 4 (G + C).

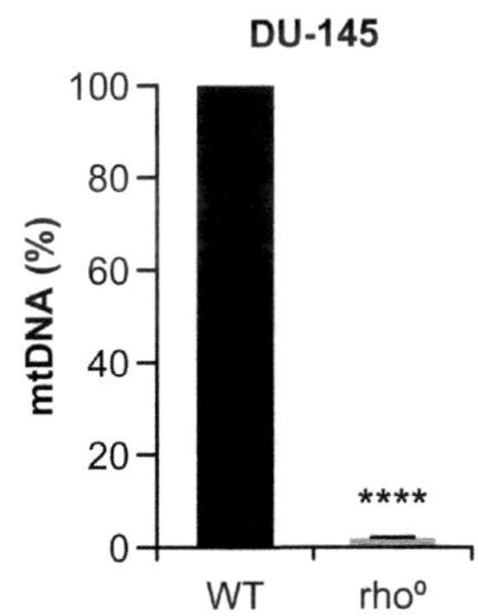

Fig. 2 mtDNA quantification. Wild-type (WT) human DU-145 prostate cancer cells were left untreated or exposed to 500 ng/mL ethidium bromide for 10 days as detailed in Section 3.1, followed by the qPCR assessment of mitochondrial DNA (mtDNA) content, as detailed in Section 3.2. Results are represented as mean ± SEM from a single experiment performed in technical duplicates. ****$P < 0.0001$ (unpaired Student's *t*-test, as compared to untreated WT cells).

3.3 Isolation of rho^0 cell clones by limiting dilution

If mtDNA can be detected after ethidium bromide treatment, we advise isolating rho^0 cell clones by limiting dilution, as follows:

1. Briefly, add 100 µL complete DMEM supplemented with 50 µg/mL uridine and 110 µg/mL pyruvate per well in a 96-well plate except well A1 (*see* ***Note 16***).
2. Seed 4×10^3 potential rho^0 cells in 200 µL complete DMEM supplemented with 50 µg/mL uridine and 110 µg/mL pyruvate in well A1.

3. Add 100 μL complete DMEM supplemented with 50 μg/mL uridine and 110 μg/mL pyruvate to wells B1–H1 to obtain a final volume of 200 μL in all wells of column 1.
4. With an 8-channel micropipette, mix and transfer 100 μL from the first column (A1–H1) to the second column (A2–H2). Repeat such a 1:2 dilution across the plate (***see Note 17***) ultimately discarding 100 μL from the last column (A12–H12) (***see Note 18***).
5. With a 12-channel micropipette, add 100 μL complete DMEM supplemented with 50 μg/mL uridine and 110 μg/mL pyruvate to wells A1–A12 to obtain a final volume of 200 μL in all wells of row A.
6. With the same micropipette, mix and transfer 100 μL from the first row (A1–A12) to the second row (B1–B12). Repeat such a 1:2 dilution across the plate (***see Note 17***) ultimately discarding 100 μL from the last row (H1–H12) (***see Note 18***).
7. Typically, isolated clones will start growing after 1–2 weeks in standard culture conditions. At confluence, passage clones into two separate wells to test proliferation in complete DMEM supplemented with 110 μg/mL pyruvate but no uridine. Collect clones that present uridine auxotrophy and analyze mtDNA abundance by qPCR (*see* Section 3.2).

3.4 Expansion of donor cells, mitochondria isolation and purification

1. Expand 22Rv1 cells with complete RPMI 1640 medium in 75 cm^2 cell culture flasks until obtaining 25×10^6 cells or more.
2. Discard exhaust medium and harvest 22Rv1 cells in a 50 mL tube after incubating them with 3 mL Trypsin/EDTA at 37 °C for 5 min, then centrifuge at $520 \times g$ for 5 min at RT (***see Note 19***).
3. Discard the supernatant, wash cells 3× with ice-cold PBS and centrifuge at $520 \times g$ for 5 min at RT (***see Note 20***).
4. Resuspend the cell pellet in ice-cold hypotonic buffer (10 mM MOPS, 83 mM sucrose, pH 7.2), with a volume 7× estimated pellet volume (e.g., 700 μL buffer for 100 μL pellet).
5. Transfer the cell suspension into a homogenizer-glass vessel and leave on ice for 2 min.
6. Disrupt cell membranes by performing between 8 and 10 strokes (***see Note 21***) in the homogenizer at 600 RPM (***see Note 22***).
7. Create an isotonic environment by adding the same volume (7× pellet volume) of ice-cold hypertonic buffer (30 mM MOPS, 250 mM sucrose, pH 7.2) to the cell suspension.

8. Mix and transfer the homogenate into a 15 mL tube.
9. Carry out 3 consecutive 5-min centrifugations at 4 °C and 1000 × *g*. After each centrifugation, transfer the supernatant (which contains mitochondria) to a new tube and discard the pellet (which contains unruptured cells and other subcellular compartments) (***see Note 23***).
10. Centrifuge at 18,000 × *g* and 4 °C for 2 min.
11. Discard the supernatant and resuspend the pellet in ice-cold buffer A.
12. Repeat this process until all mitochondria are sedimented in a single tube (***see Notes 24 and 25***). Keep mitochondrial extracts in ice.

3.5 Mitochondrial Transfer (*see Note 22*)

1. Resuspend 2–3 × 10^6 rho^0 cells in 2 mL complete DMEM supplemented with 50 μg/mL uridine and 110 μg/mL pyruvate.
2. Carefully lay the rho^0 cell suspension on top of the mitochondria pellet and centrifuge at 520 × *g* for 5 min to allow cells to mix with mitochondria.
3. Remove the supernatant and add 100 μL PEG solution (50%), gently resuspend the pellet for 30 s and allow to rest untouched for another 30 s. This process facilitates the incorporation of mitochondria by cells.
4. Transfer the mixture into a 6-well plate with complete DMEM supplemented with 110 μg/mL pyruvate (but no uridine) and incubate at 37 °C and 5% CO_2. After 5–7 days, rho^0 cells will succumb to the absence of uridine, enabling for the selection of transmitochondrial cybrids.

3.6 Analysis of mitochondrial purity

To verify the degree of heteroplasmy in transmitochondrial cybrids, restriction fragment length polymorphism (RFLP) based on a restriction enzyme that specifically recognizes a mtDNA palindrome from donor (but not recipient) cells is employed (***see Note 26***), as follows:

1. Expand 1–5 × 10^6 transmitochondrial cybrids in with complete DMEM supplemented with 110 μg/mL pyruvate and collect them upon detachment with 3 mL Trypsin/EDTA, and centrifugation at 520 × *g* for 5 min.
2. Discard supernatant and resuspend the cell pellet in 5 mL PBS.
3. Extract DNA from the sample as described in Section 3.2.1.
4. Use 100 ng/mL DNA template in a standard PCR thermal protocol that, in the case of DU-145 and 22Rv1 cells, aims at generating a 1233-bp amplicon based on 5′-GTGCCACCTATCACACCCCA-3′ and 5′-GAGTGGGGTTTTGCAGTCCTT-3′ forward and reverse primers (Table 3).

Table 3 Thermal protocol for analysis of mitochondrial purity by PCR.

Step[a]	**Time**	**Temperature**
Initial denaturation	2 min	95 °C
Denaturation	30 s	95 °C
Annealing	30 s	Chosen according to the T_m of the primers used[b]
Extension	1–2 min[c]	72 °C
Final extension	5 min	72 °C

[a]Steps 2–4 were repeated for 29 cycles.
[b]This temperature depends on the primer sequence and length, and it can be calculated by using the following equation: Temperature = 2 (A + T) + 4 (G + C).
[c]This time depending on the length of the amplicons: for fragments >1 kbp 2 min are recommended.

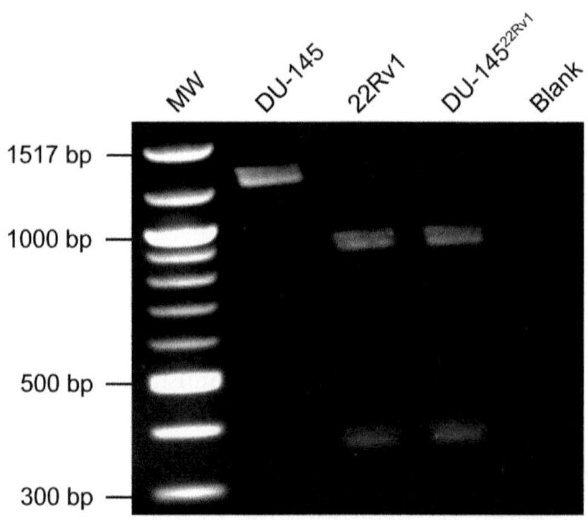

Fig. 3 Mitochondrial purity verification. DNA was isolated from wild-type (WT) human DU-145 prostate cancer cells, WT human 22Rv1 prostate cancer cells and DU-145^{22Rv1} cybrids as detailed in Section 3.2.1, followed by PCR for the amplification of a mitochondrial amplicon of 1233 bp, enzymatic digestion with *SacI*-HF® and product resolution on a 1.5% agarose gel, as detailed in Section 3.6.

5. On ice, mix 1 μg amplified DNA, 5 μL 10× NEBuffer as provided by restriction enzyme kit, 1 μL *SacI*-HF® and nuclease-free water up to 50 μL total volume.
6. Gently mix by pipetting, and microfuge briefly at 500 g for 1 min.
7. Incubate at 37 °C for 15 min (***see Note 27***) and at 65 °C for 20 min to inactivate restriction enzymes.
8. Separate DNA fragments in a 1.5% agarose gel.
9. Check the fragment pattern by exposing under UV light in the Azure 600® Imager (Fig. 3).

4. Notes

1. Virtually equivalent disposables, reagents, equipment and software are available from a variety of alternative providers.
2. Ethidium bromide is a strong irritant as well as a powerful mutagenic agent, and hence should be handled with appropriate safety precautions.
3. This protocol is precisely tailored to DU-145 cells as recipient cells and 22Rv1 cells as mitochondrial donor cells, but minimal variations should be sufficient for it to be applicable to other cancer cell lines.
4. To avoid contaminations, the entire protocol should be performed under sterile conditions and with sterile (filtered or autoclaved) reagents.
5. Uridine and pyruvate are required to support the growth of mtDNA-depleted rho^0 cells.
6. Cells are routinely grown at 37 °C under 5% CO_2 in small (25 cm^2) or medium (75 cm^2) culture flasks.
7. Some cell lines might require alternative ethidium bromide dosing and/or administration schedules.
8. If mtDNA can still be detected, rho^0 clones can be isolated from the cell population by limiting dilution (see Section 3.3).
9. Some cell lines might develop resistance to ethidium bromide, which (at least in some cases) can be circumvented by increasing ethidium bromide concentration and shortening treatment duration.
10. Any qPCR primers that specifically recognize nuclear and mitochondrial genes can be used for this analysis. In this case, we designed qPCR primers with Primer Express 2.0 software tested them functionally in quantitative analysis.
11. DNA extracted from untreated cells provides a convenient positive control, pure water a convenient negative one.
12. Buffer AE can be replaced by distilled water.
13. DNA samples can be safely stored at −20 °C for 12–24 months in the absence of sizeable degradation.
14. Each sample should be analyzed for the abundance of both nuclear and mitochondrial amplicons.
15. Keep capillaries set in an ice-cold aluminum block.
16. Conditioned media from subconfluent DU-145 cell cultures can be employed to support the survival of single clones.
17. It is recommended to avoid the formation of bubbles during pipetting.
18. At this stage, all wells should contain 100 μL total volume.

19. Optimal collection strategy and centrifugation conditions may vary across different cell lines (e.g., trypsinization is not necessary for cells growing in suspension).
20. All steps for mitochondrial extraction and purification should be carried out with ice-cold reagents and ice-cold tubes.
21. Stroke refers to the action of pulling down the homogenizer glass tube to create a temporary vacuum while the pestle is rotating.
22. Cell membranes composition and strength vary across cell lines, implying that optimization may be required.
23. To avoid contamination with other cellular components or intact cells, supernatant should be collected as carefully as possible and only up to 75% of total volume, with the objective of leaving the pellet undisturbed.
24. This step ensures the precipitation of isolated mitochondria.
25. Mitochondrial purity can be quickly evaluated under light microscopy upon incubation of a small aliquot with 0.4% trypan blue in PBS (which marks nuclei, both inside and outside cells), 1:1 (v:v).
26. The restriction enzyme used for RFLP analysis strictly depends on donor cells. Here, we employed *SacI*-HF®, which specifically recognizes a palindrome in the mtDNA of 22Rv1 cells.
27. Restriction enzymes that display no unspecific (star) activity can be safely allowed to operate overnight.

5. Concluding remarks

Here, we provide a simple and fairly inexpensive method for generating transmitochondrial cybrids in cancer cells, based on (1) the establishment of mtDNA-depleted rho^0 cells as recipient cells, (2) the extraction of intact mitochondria from donor cells, and (3) the delivery of mitochondria from donor cells to recipient cells. This approach has been successfully used to establish human prostate cancer DU-145 cells bearing mitochondria from human prostate cancer 22Rv1 cells (Soler-Agesta et al., 2023) for the characterization of the mode of action of a novel immunotherapeutic agent known as PT-112 (Karp et al., 2022; Soler-Agesta et al., 2022; Yamazaki et al., 2020b), as well as mouse WT L929 fibroblasts bearing mitochondria from a subline thereof with a detached phenotype (L929dt cells), for the study of mitochondrial metabolism (Catalán et al., 2015; Marchi et al., 2020; Marco-Brualla et al., 2019).

While the protocol presented herein is specifically tailored to DU-145 recipient and 22Rv1 donor cells, we expect it to be simply adaptable to other (cancer) cell lines with minor experimental variations. One of the limitations of this method emerges from the ability of some cell lines to acquire resistance to ethidium bromide. While in our settings simply increasing ethidium bromide concentration and shortening treatment duration was sufficient to circumvent such issue, an alternative solution consists in combining ethidium bromide with dideoxycytidine (ddC), a reverse transcriptase inhibitor that has been shown to inhibit mtDNA replication (Nelson et al., 1997). Moreover, suboptimal mtDNA depletion may be circumvented by isolating truly rho^0 clones via limiting dilution. In this latter scenario, however, subsequent experimental assays should be performed on multiple independent control and rho^0 clones to account for potential clone-dependent mtDNA-independent effects. Despite these and other caveats, the protocol detailed herein stands out as a simple approach to study the influence of specific mtDNA genotypes on cancer cell functions, both in vitro and in vivo.

Acknowledgments

This work was partially supported the Spanish Association Against Cancer in Aragón (#PRDAR21487SOLE), proving financial support to R.S.A.. C.R.Y. work is supported by the European Union-NextGenerationEU. The LG lab is/has been supported (as a PI unless otherwise indicated) by one NIH R01 grant (#CA271915), by two Breakthrough Level 2 grants from the US DoD BCRP (#BC180476P1, #BC210945), by a grant from the STARR Cancer Consortium (#I16-0064), by a Transformative Breast Cancer Consortium Grant from the US DoD BCRP (#W81XWH2120034, PI: Formenti), by a U54 grant from NIH/NCI (#CA274291, PI: Deasy, Formenti, Weichselbaum), by the 2019 Laura Ziskin Prize in Translational Research (#ZP-6177, PI: Formenti) from the Stand Up to Cancer (SU2C), by a Mantle Cell Lymphoma Research Initiative (MCL-RI, PI: Chen-Kiang) grant from the Leukemia and Lymphoma Society (LLS), by a Rapid Response Grant from the Functional Genomics Initiative (New York, US), by a pre-SPORE grant (PI: Demaria, Formenti), a Collaborative Research Initiative Grant and a Clinical Trials Innovation Grant from the Sandra and Edward Meyer Cancer Center (New York, US), by startup funds from the Dept. of Radiation Oncology at Weill Cornell Medicine (New York, US), by industrial collaborations with Lytix Biopharma (Oslo, Norway), Promontory (New York, US) and Onxeo (Paris, France), as well as by donations from Promontory (New York, US), the Luke Heller TECPR2 Foundation (Boston, US), Sotio a.s. (Prague, Czech Republic), Lytix Biopharma (Oslo, Norway), Onxeo (Paris, France), Ricerchiamo (Brescia, Italy), and Noxopharm (Chatswood, Australia).

Competing interests

L.G. is/has been holding research contracts with Lytix Biopharma, Promontory and Onxeo, has received consulting/advisory honoraria from Boehringer Ingelheim, AstraZeneca, AbbVie, OmniSEQ, Onxeo, The Longevity Labs, Inzen, Imvax, Sotio, Promontory, Noxopharm, EduCom, and the Luke Heller TECPR2 Foundation, and holds Promontory stock options. The other authors have no conflicts of interest to declare.

References

Catalán, E., et al. (2015). MHC-I modulation due to changes in tumor cell metabolism regulates tumor sensitivity to CTL and NK cells. *Oncoimmunology, 4*, e985924. https://doi.org/10.4161/2162402x.2014.985924.

Cavaliere, A., et al. (2022). An in vitro approach to study mitochondrial dysfunction: A cybrid model. *Journal of Visualized Experiments*. https://doi.org/10.3791/63452.

Chavez, M., et al. (2023). Advances in CRISPR therapeutics. *Nature Reviews. Nephrology, 19*, 9–22. https://doi.org/10.1038/s41581-022-00636-2.

Craven, L., et al. (2017). Recent advances in mitochondrial disease. *Annual Review of Genomics and Human Genetics, 18*, 257–275. https://doi.org/10.1146/annurev-genom-091416-035426.

DeBerardinis, R. J., & Chandel, N. S. (2020). We need to talk about the Warburg effect. *Nature Metabolism, 2*, 127–129. https://doi.org/10.1038/s42255-020-0172-2.

Diepstraten, S. T., et al. (2022). The manipulation of apoptosis for cancer therapy using BH3-mimetic drugs. *Nature Reviews. Cancer, 22*, 45–64. https://doi.org/10.1038/s41568-021-00407-4.

Dora, D., et al. (2023). Non-small cell lung cancer patients treated with anti-PD1 immunotherapy show distinct microbial signatures and metabolic pathways according to progression-free survival and PD-L1 status. *Oncoimmunology, 12*, 2204746. https://doi.org/10.1080/2162402X.2023.2204746.

Gustafsson, C. M., et al. (2016). Maintenance and expression of mammalian mitochondrial DNA. *Annual Review of Biochemistry, 85*, 133–160. https://doi.org/10.1146/annurev-biochem-060815-014402.

Hanahan, D. (2022). Hallmarks of cancer: New dimensions. *Cancer Discovery, 12*, 31–46. https://doi.org/10.1158/2159-8290.Cd-21-1059.

Hanahan, D., & Weinberg, R. A. (2011). Hallmarks of cancer: The next generation. *Cell, 144*, 646–674. https://doi.org/10.1016/j.cell.2011.02.013.

Karp, D. D., et al. (2022). Phase I study of PT-112, a novel pyrophosphate-platinum immunogenic cell death inducer, in advanced solid tumours. *EClinicalMedicine, 49*, 101430. https://doi.org/10.1016/j.eclinm.2022.101430.

Kopinski, P. K., et al. (2021). Mitochondrial DNA variation and cancer. *Nature Reviews. Cancer, 21*, 431–445. https://doi.org/10.1038/s41568-021-00358-w.

Koppenol, W. H., et al. (2011). Otto Warburg's contributions to current concepts of cancer metabolism. *Nature Reviews. Cancer, 11*, 325–337. https://doi.org/10.1038/nrc3038.

Kroemer, G., et al. (2023). Immunosurveillance in clinical cancer management. *CA: a Cancer Journal for Clinicians*. https://doi.org/10.3322/caac.21818.

Locasale, J. W., et al. (2011). Phosphoglycerate dehydrogenase diverts glycolytic flux and contributes to oncogenesis. *Nature Genetics, 43*, 869–874. https://doi.org/10.1038/ng.890.

Lunt, S. Y., & Vander Heiden, M. G. (2011). Aerobic glycolysis: Meeting the metabolic requirements of cell proliferation. *Annual Review of Cell and Developmental Biology, 27*, 441–464. https://doi.org/10.1146/annurev-cellbio-092910-154237.

Madigan, V., et al. (2023). Drug delivery systems for CRISPR-based genome editors. *Nature Reviews. Drug Discovery, 22*, 875–894. https://doi.org/10.1038/s41573-023-00762-x.

Marchi, S., et al. (2020). Ca(2+) fluxes and cancer. *Molecular Cell, 78*, 1055–1069. https://doi.org/10.1016/j.molcel.2020.04.017.

Marchi, S., et al. (2023). Mitochondrial control of inflammation. *Nature Reviews. Immunology, 23*, 159–173. https://doi.org/10.1038/s41577-022-00760-x.

Marco-Brualla, J., et al. (2019). Mutations in the ND2 subunit of mitochondrial complex I are sufficient to confer increased tumorigenic and metastatic potential to cancer cells. *Cancers (Basel), 11*. https://doi.org/10.3390/cancers11071027.

Medini, H., et al. (2020). Mitochondria are fundamental for the emergence of metazoans: On metabolism, genomic regulation, and the birth of complex organisms. *Annual Review of Genetics, 54*, 151–166. https://doi.org/10.1146/annurev-genet-021920-105545.

Nelson, I., et al. (1997). Depletion of mitochondrial DNA by ddC in untransformed human cell lines. *Somatic Cell and Molecular Genetics*, *23*, 287–290. https://doi.org/10.1007/BF02674419.

Neupert, W. (2016). Mitochondrial gene expression: A playground of evolutionary tinkering. *Annual Review of Biochemistry*, *85*, 65–76. https://doi.org/10.1146/annurev-biochem-011116-110824.

Newman, L. E., & Shadel, G. S. (2023). Mitochondrial DNA release in innate immune signaling. *Annual Review of Biochemistry*, *92*, 299–332. https://doi.org/10.1146/annurev-biochem-032620-104401.

Oldan, J. D., et al. (2023). Increased tryptophan, but not increased glucose metabolism, predict resistance of pembrolizumab in stage III/IV melanoma. *Oncoimmunology*, *12*, 2204753. https://doi.org/10.1080/2162402X.2023.2204753.

Petroni, G., et al. (2021). Immunomodulation by targeted anticancer agents. *Cancer Cell*, *39*, 310–345. https://doi.org/10.1016/j.ccell.2020.11.009.

Porporato, P. E., et al. (2018). Mitochondrial metabolism and cancer. *Cell Research*, *28*, 265–280. https://doi.org/10.1038/cr.2017.155.

Rackham, O., & Filipovska, A. (2022). Organization and expression of the mammalian mitochondrial genome. *Nature Reviews. Genetics*, *23*, 606–623. https://doi.org/10.1038/s41576-022-00480-x.

Roy, A., et al. (2022). Mitochondrial DNA replication and repair defects: Clinical phenotypes and therapeutic interventions. *Biochimica et Biophysica Acta - Bioenergetics*, *1863*, 148554. https://doi.org/10.1016/j.bbabio.2022.148554.

Schwenck, J., et al. (2023). Advances in PET imaging of cancer. *Nature Reviews. Cancer*, *23*, 474–490. https://doi.org/10.1038/s41568-023-00576-4.

Soler-Agesta, R., et al. (2022). PT-112 induces mitochondrial stress and immunogenic cell death, targeting tumor cells with mitochondrial deficiencies. *Cancers (Basel)*, *14*. https://doi.org/10.3390/cancers14163851.

Soler-Agesta, R., et al. (2023). Transmitochondrial cybrid generation using cancer cell lines. *Journal of Visualized Experiments*. https://doi.org/10.3791/65186.

Sprooten, J., et al. (2023). Trial watch: Chemotherapy-induced immunogenic cell death in oncology. *Oncoimmunology*, *12*, 2219591. https://doi.org/10.1080/2162402X.2023.2219591.

Stewart, J. B., & Chinnery, P. F. (2021). Extreme heterogeneity of human mitochondrial DNA from organelles to populations. *Nature Reviews. Genetics*, *22*, 106–118. https://doi.org/10.1038/s41576-020-00284-x.

Stine, Z. E., et al. (2022). Targeting cancer metabolism in the era of precision oncology. *Nature Reviews. Drug Discovery*, *21*, 141–162. https://doi.org/10.1038/s41573-021-00339-6.

Thompson, C. B., et al. (2023). A century of the Warburg effect. *Nature Metabolism*, *5*, 1840–1843. https://doi.org/10.1038/s42255-023-00927-3.

Vander Heiden, M. G., et al. (2009). Understanding the Warburg effect: The metabolic requirements of cell proliferation. *Science*, *324*, 1029–1033. https://doi.org/10.1126/science.1160809.

Vaupel, P., & Multhoff, G. (2021). Revisiting the Warburg effect: Historical dogma versus current understanding. *The Journal of Physiology*, *599*, 1745–1757. https://doi.org/10.1113/jp278810.

Villiger, L., et al. (2024). CRISPR technologies for genome, epigenome and transcriptome editing. *Nature Reviews. Molecular Cell Biology*. https://doi.org/10.1038/s41580-023-00697-6.

Vitale, I., et al. (2023). Apoptotic cell death in disease-current understanding of the NCCD 2023. *Cell Death and Differentiation*, *30*, 1097–1154. https://doi.org/10.1038/s41418-023-01153-w.

Vringer, E., & Tait, S. W. G. (2023). Mitochondria and cell death-associated inflammation. *Cell Death and Differentiation*, *30*, 304–312. https://doi.org/10.1038/s41418-022-01094-w.

Vyas, S., et al. (2016). Mitochondria and Cancer. *Cell*, *166*, 555–566. https://doi.org/10.1016/j.cell.2016.07.002.

Warburg, O. (1956a). On respiratory impairment in cancer cells. *Science*, *124*, 269–270.

Warburg, O. (1956b). On the origin of cancer cells. *Science*, *123*, 309–314. https://doi.org/10.1126/science.123.3191.309.

Warburg, O., & Minami, S. (1923). Versuche an Überlebendem Carcinom-gewebe. *Klinische Wochenschrift*, *2*, 776–777. https://doi.org/10.1007/BF01712130.

Warburg, O., et al. (1924). Über den stoffwechsel der carcinomzelle. *Naturwissenschaften*, *12*, 1131–1137.

Yamazaki, T., et al. (2020a). Mitochondrial DNA drives abscopal responses to radiation that are inhibited by autophagy. *Nature Immunology*, *21*, 1160–1171. https://doi.org/10.1038/s41590-020-0751-0.

Yamazaki, T., et al. (2020b). PT-112 induces immunogenic cell death and synergizes with immune checkpoint blockers in mouse tumor models. *Oncoimmunology*, *9*, 1721810. https://doi.org/10.1080/2162402x.2020.1721810.

Yang, X., et al. (2019). Mitochondrial DNA mutation, diseases, and nutrient-regulated Mitophagy. *Annual Review of Nutrition*, *39*, 201–226. https://doi.org/10.1146/annurev-nutr-082018-124643.

Yankeelov, T. E., et al. (2014). Quantitative multimodality imaging in cancer research and therapy. *Nature Reviews. Clinical Oncology*, *11*, 670–680. https://doi.org/10.1038/nrclinonc.2014.134.

Yuan, J., & Ofengeim, D. (2023). A guide to cell death pathways. *Nature Reviews. Molecular Cell Biology*. https://doi.org/10.1038/s41580-023-00689-6.

Zhang, H., et al. (2021). Perfecting targeting in CRISPR. *Annual Review of Genetics*, *55*, 453–477. https://doi.org/10.1146/annurev-genet-071719-030438.

Zong, W. X., et al. (2016). Mitochondria and Cancer. *Molecular Cell*, *61*, 667–676. https://doi.org/10.1016/j.molcel.2016.02.011.

CHAPTER THREE

Enzyme-linked ImmunoSpot (ELISpot) assay to quantify peptide-specific IFN-γ production by splenocytes in a mouse tumor model after radiation therapy

Benoît Lecoester[a], Yaoyao Xie[a], Amélie Marguier[a], Laura Boulerot[a], Marine Malfroy[a], Olivier Adotévi[a,b], and Jihane Boustani[a,c,*]

[a]INSERM, EFS BFC, UMR 1098, RIGHT, Greffon-Hôte-Tumeur Interactions/Ingénierie Cellulaire et Génique, Université de Franche-Comté, Besançon, France
[b]Department of Medical Oncology, University Hospital of Besançon, Besançon, France
[c]Department of Radiation Oncology, University Hospital of Besançon, Besançon, France
*Corresponding author: e-mail address: jboustani@chu-besancon.fr

Contents

Methods in Cell Biology, Volume 189
ISSN 0091-679X
https://doi.org/10.1016/bs.mcb.2024.07.001

Abstract

To develop new effective therapeutic strategies for cancer patients, there is a need for extensive and precise insights into the mechanisms involved in the immune response to anti-cancer treatments. The enzyme-linked immunospot (ELISpot) assay is a rapid and reproducible technique that allows the detection of cytokine-producing antigen-specific T cells at the single cell level. This protocol describes an interferon gamma (IFN-γ) ELISpot method for measuring antigen-specific murine CD8$^+$ T cells that produce IFN-γ, a marker of their activation and cytotoxicity. Splenocytes from tumor-bearing mice treated with radiation therapy were used as source of CD8$^+$ T cells and were stimulated with a tumor-derived peptide. This method was facilitated by a ready-to-use assay kit and provides a tool to analyze the specificity, intensity, and kinetics of specific CD8$^+$ T cells.

1. Introduction

There is a widespread desire for knowledge about the antigenic specificity and cytokine secretion potential of T cells in several fields such as cancer, autoimmune diseases, infectiology, and vaccinology (Laheurte et al., 2019; Möbs & Schmidt, 2016). Mice-based experiments provide an interesting in vivo model to study T cell functions. Although several differences can be observed, the immune cells and the mechanisms involving them are similar to those found in humans (Masopust, Sivula, & Jameson, 2017). Therefore, multiple mouse models of cancer were developed to better understand the effect of anti-cancer treatments, such as chemotherapy, immunotherapy, and radiation therapy. One of the objectives of these treatments is to re-establish the antitumor immunity mediated by T cells. When polarized as T_H1 cells, CD4$^+$ T lymphocytes coordinate the immune response and stimulate a pro-inflammatory tumor microenvironment (Borst, Ahrends, Bąbała, Melief, & Kastenmüller, 2018). Triggering a CD4$^+$ T anti-tumor response is particularly relevant for enhancing the cytotoxic T response of CD8$^+$ T cells and stimulating memory response (Ahrends et al., 2019). The study of CD8$^+$ T cells within the spleen appears to be a good way of monitoring the peripheral immune response in treated mice. Here, we propose to analyze the prevalence and intensity of the tumor-specific CD8$^+$ T cell response in the spleen of tumor-bearing mice tumors treated with radiation therapy. First, we injected subcutaneously colon cancer cells into the left flank of mice. Once they became macroscopically visible, the tumors were irradiated. Seven days following treatment, spleens were collected to isolate the splenocytes that primarily comprise T lymphocytes (Hensel, Khattar, Ashton, & Ponnazhagan, 2019). Finally, we performed an ELISpot assay to measure the secretion of interferon gamma (IFN-γ), a key cytokine of T_H1 polarization after Major Histocompatibility Class I-restricted tumor-derived peptide stimulation.

2. Materials

2.1 Disposables

- Malassez Counting Chamber (e.g., #DHC-M01, NanoEntech Inc., South Korea) (**see Note 1**)
- Sterile needle 25G for subcutaneous injection (e.g., #300-600, BD Biosciences, Franklin Lakes, New Jersey, USA) **(see Note 1)**
- Sterile needle 20G for intraperitoneal injection (e.g., #301-300, BD Biosciences) **(see Note 1)**
- Sterile syringe 1 mL for tumor and drug injection (#Omnifix-F-916-1406 V, Braun™, Melsungen, Deutschland) **(see Note 1)**
- Sterile 96-well plates (e.g., #353072, Corning Incorporated, Durham, North Carolina, USA) **(see Note 1)**
- Sterile 6-well plates (e.g., #353046, Corning Incorporated) **(see Note 1)**
- Sterile 175 cm^2 tissue culture flasks with vanted cap (e.g., #353112, Corning Incorporated) **(see Note 1)**
- Sterile 50 mL Falcon® Polypropylene Tubes (e.g., #352070, Corning Sciences, Mexico, Mexico) **(see Note 1)**
- 0.5 mL Eppendorf tubes (e.g., #0030124537, Fisher Scientific, Bernolsheim, France) **(see Note 1)**
- Sterile Serological Pipets with volume 5, 10 or 25 mL (e.g., ##4488 Stripette Serological Pipets 10 mL, Corning Incorporated) **(see Note 1)**
- Sterile disposable pipette tips (for 1 mL, 200 µL and 10 µL, TipOne, Starlab, Orsay, France) **(see Note 1)**
- Pipet-Aid (as an example #075212, Drummond, Dutscher, Bernolsheim, France) **(see Note 1)**
- Sterile injection syringes (e.g., #309649, 5 mL, BD Biosciences) **(see Note 1)**
- Sterile needles 27 G (e.g., #300635 BD Microlance®, BD Biosciences) **(see Note 1)**
- Modified Alcohol (e.g., Alcool modifié, Cooper, Melun, France) **(see Note 1)**

- Syringe pusher (e.g., pusher of 2 mL Injekt® Luer Solo, Braun™) **(see Note 1)**
- Sterile 100 μm filter (e.g., #352360 Falcon™ Cell Strainers, ThermoFisher, Waltham, Massachusetts, USA) **(see Note 1)**
- Murine IFN-γ ELISpot Set with non-sterile (#862.031.001P) or sterile plates (#862.031.001S) (Diaclone, Besançon, France) **(see Note 1)**
- 600 mL Beaker **(see Note 1)**
- Gloves **(see Note 1)**

2.2 Mice, cell line

- CT26 murine colon cancer cell line (e.g., CRL-2638™, ATCC, Manassas, Virginia, USA) **(see Note 1)**
- BALB/c mice (e.g., Charles River, Wilmington, Delaware, USA) **(see Notes 1 and 2)**

2.3 Reagents

- RPMI Medium 1640 + GlutaMAX-I (e.g., #72400-021, ThermoFisher) **(see Note 1)**
- Heat-inactivated fetal bovine serum (FBS)
- Penicillin-Streptomycin (PS) Solution (e.g., #15140130, ThermoFisher) **(see Note 1)**
- X-Vivo 15 (e.g., #BESP1230Q, Lonza, Verviers, Belgium) **(see Note 1)**
- Phosphate Buffered Saline (PBS) pH 7,2 (1×) (e.g., #20012-019, ThermoFisher) **(see Note 1)**
- Trypsin 0.5%/EDTA 0.2% in DPBS, w/o: Ca and Mg (e.g., #P10-023100, Pan Biotech, Aidenbach, Deutchland) **(see Note 1)**
- Trypan Blue Solution, 0.4% (#15250061, ThermoFisher) **(see Note 1)**
- Tween 20 (e.g., #003005, ThermoFisher) **(see Note 1)**
- Ethanol
- Distilled Water
- PMA-ionomycin (e.g., #00-4970-93, ThermoFisher) **(see Note 1)**
- MHC class I restricted tumor-derived peptides (e.g., AH1 peptides, gp70 H2-L^d-restricted epitope, #AH11-P-1, Life Technologies™, Delhi, India) **(see Note 1)**

2.4 Equipments

- Humidified cell culture incubators at 37 °C with 5% CO^2
- Laboratory safety cabinet (Class II)
- Optical microscope

- Single and multichannel pipettes
- Water bath
- Lab bench centrifuge (e.g., Rotina 420/420 R, Hettich France Sarl, France) **(see Note 1)**
- Animal cage cabinet
- Compact anesthesia unit (Minerve, Esternay, France) **(see Note 1)**
- Digital calipers (MOIN1150MI, VWR, Rosny-sous-Bois, France) **(see Note 1)**
- Material for irradiation **(See Ref:** Mirjolet, Ladjohounlou, Baude, & Bessières, 2023**)**
- ELISpot plates analyzer (e.g., ImmunoSpot Series 6 Analyzer, Cellular Technology Limited, Ohio, USA) **(see Note 1)**
- ELISpot analysis software (e.g., ImmunoSpot 6.0 Software, Cellular Technology Limited) **(see Note 1)**

3. Experiment steps

Example of cell culture for the subcutaneous graft of CT26 cells onto mice.

3.1 Cell preparation

3.1.1 Defrost

- Remove FBS, RPMI 1640, and PS from the fridge (ideally thawed 24 h beforehand).
- Remove the freezing tube containing the CT26 cell line from the stock freezer.
- Defrost cells. The use of a 37 °C water bath is mandatory.
- Transfer, as soon as defrosting allows, the defrozed cells in 5 mL of RPMI 1640 medium supplemented with 10% FBS and 1% PS in 15 mL tube.
- Rinse with RPMI 10% FBS 1% PS the freezing tube thoroughly to recover all cells. Adjust volume to 10 mL.
- Centrifugation 1500 RPM for 5 min at room temperature (RT).
- Resuspend pellet in 10 mL RPMI 10% FBS 1% PS.
- Transfer 30 µL of the suspension into a 96-well plate and dilute it with 30 µL of trypan blue (1/2 dilution). Count cells using a Malassez cell counter.
- Transfer an adjusted concentration of cells to the flask you wish to use.
- CT26 cell line is incubated in 75 cm^2 flask in RPMI 1640 medium supplemented with 10% FBS and 1% PS at 37 °C in a humidified atmosphere containing 5% CO_2 **(see Note 3)**.

3.1.2 Cell passage

- Check cells in the flask under a microscope to confirm that the cells are 80–90% confluent. If this is the case, you can carry out the cell passage.
- Discard the culture medium from the adherent flask and wash the flask with 2 mL of PBS 1×.
- Harvest the cells (with approximately 80–90% confluency) in a 15 mL tube upon incubation with 1 mL trypsin-EDTA 0.05% during 3 min at 37 °C. Check that the trypsin is in contact with the entire cell surface of the flask. Add RPMI 1640 + 10% FBS+ 1% PS per flask (always adjust according to flask size) after 3 min of trypsin-EDTA to facilitate cell transfer from the flask to the 15 mL tube.
- Adjust to 10 mL of RPMI 1640 medium in 15 mL tube.
- Centrifugate 1500 rpm for 5 min at RT.
- Resuspend cell pellets with 10 mL of RPMI 1640 medium.
- Transfer 30 µL of the suspension into a 96-well plate and dilute it with 30 µL of trypan blue (1/2 dilution). Count cells using a Malassez cell counter.
- Transfer an adjusted concentration of cells to the flask you wish to use.
- CT26 cell line is incubated in a 75 cm^2 flask in RPMI 1640 medium supplemented with 10% FBS and 1% PS at 37 °C in a humidified atmosphere containing 5% CO_2 **(See Note 4)**.

3.2 Engraftment

- The day before tumor transplantation, shave the mice on the engraftment flank to prepare for cancer cell inoculation.
- On the engraftment day, prepare cells in order to obtain enough to inject all individuals. Each mouse will receive an injection of 2.10^5 CT26 cells in 100 µL of PBS 1× **(see Note 5)**.
- Anesthetize the mouse in the induction cage of the Minerve system with 2.5% isoflurane mixed with medical pure oxygen for 2–3 min.
- Proceed to the subcutaneous injection of CT26 cells into mice at the shaved area.

3.3 Treatment

When the tumor reaches a macroscopically visible volume **(See Note 6)** you can introduce radiation therapy, the protocol of which is described in: «Novel platform for subcutaneous tumor irradiation in mice», Methods in Cell Biology, Academic Press, 2023 (Mirjolet et al., 2023).

3.4 Splenocyte isolation

- Prepare the work surface with 2 forceps and a sterile scalpel.
- Sacrifice the mouse by carbon dioxide or cervical dislocation euthanasia.
- Soak the mouse completely in alcohol before placing it on a dissecting board.
- Dissect under sterile conditions.
- Make an incision along the midline of the abdomen. Use tacks to hold the abdominal skin taut and allow vision and access to the abdomen (Fig. 1).
- Identify and remove the spleen from the mouse's left hypochondrium (Fig. 1).
- Add 3 mL of RPMI 1640 medium at RT to each well in a 6-well plate (as many wells as spleens to be harvested) (Fig. 2).
- Add a filter to the top of the wells, then transfer the spleen (Fig. 2).
- Crush each spleen, then rinse the filter **(See Note 7)** (Fig. 2).
- Remove the filter and transfer the filtered cell solution to a 50 mL tube.
- Adjust the volume to 20 mL.
- Centrifugate 1500 RPM for 5 min at RT.
- Discard supernatant.
- Add 1 mL ACK Lysis solution for 3 min.
- Stop the lysis reaction by adding 19 mL RPMI.
- Centrifugate 1500 RPM for 5 min at RT.

A B C

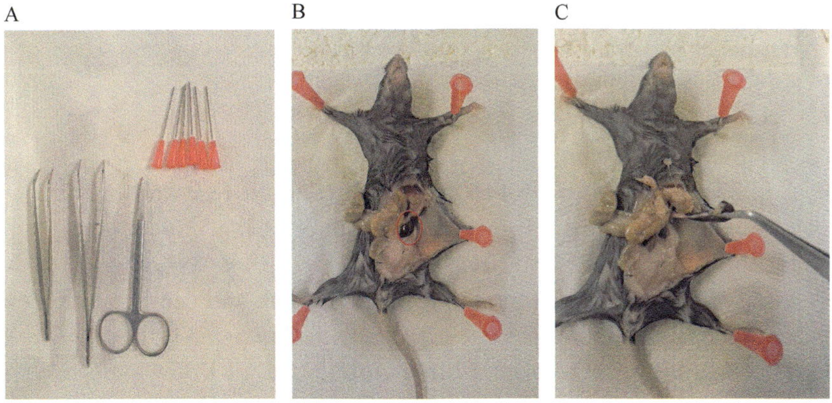

Fig. 1 Dissection and extraction of the spleen. (A) Materials required for the dissection. (B) Open the abdomen with a longitudinal midline incision and tension the anterior wall on the left side using a tack. (C) Careful extraction of the spleen from the mouse's left hypochondrium.

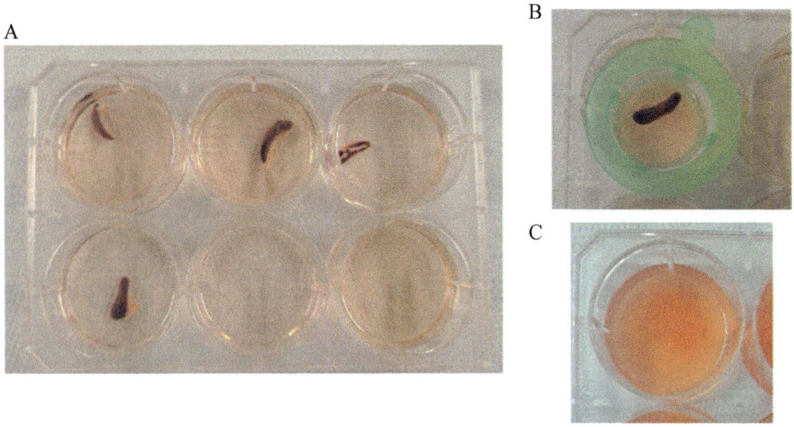

Fig. 2 Manual grinding of the spleen to obtain a splenocyte suspension. (A) Extracted spleens in a 6-well plate prior to the grinding step. (B) Using a 100-μm filter to grind spleen. (C) Splenocytes' suspension after manual grinding with a syringe plunger.

- Discard the supernatant and resuspend the cell pellet in 20 mL.
- Transfer 10 μL of the suspension into a 96-well plate and dilute it with 40 μL of PBS 1× (1/5 dilution) and re-dilute 30 μL of this 1/5 solution in 30 μL of trypan blue (1/10 final dilution). Count cells using a Malassez cell counter **(See Note 8)**.
- Transfer 5.10^6 splenocytes into 5 mL tubes.
- Centrifugate 1500 RPM for 5 min at RT.
- Resuspend the cell pellet with 1 mL of X-VIVO 15 to obtain a concentration of 5.10^6/mL.

3.5 ELISpot

3.5.1 Plate preparation

- Add 25 μL of 35% ethanol to every well **(See Note 9 and** Fig. 3).

  For one plate, mix 3.5 mL of ethanol with 6.5 mL of distilled water.
- Incubate plate at RT for 30 s.
- Empty the wells by flicking the plate over a sink and gently tapping on absorbent paper.
- Thoroughly wash the plate 5 times with 250 μL of PBS 1× per well.
- Add 100 μL of diluted capture antibody to every well.

  For one plate, dilute 100 μL of capture antibody in 10 mL of PBS 1× and mix well.
- Cover the plate and incubate at 4 °C overnight.

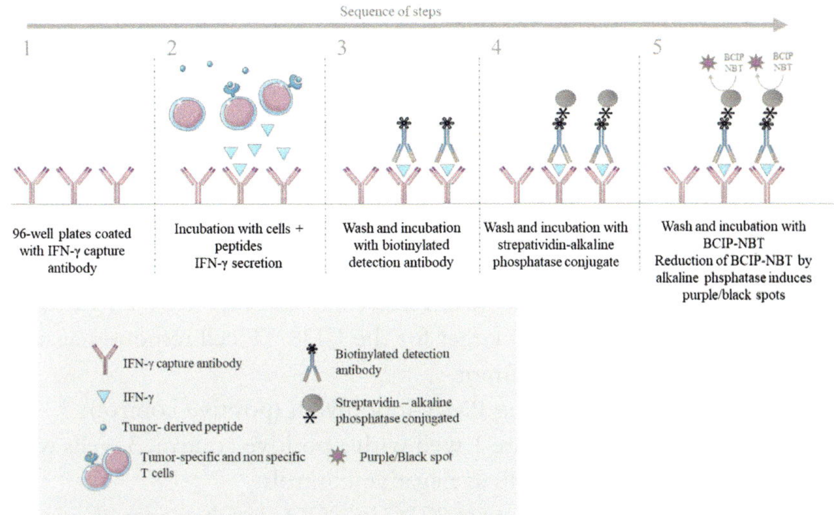

Fig. 3 ELISpot assay procedure for the detection of tumor-specific TH1-polarized T cells. (1) A capture antibody highly specific for IFN-γ is deposited in a 96-well plate. The plate is then blocked to minimize antibody-dependent non-specific binding, and washed. (2) The cell suspension containing splenocytes is added, along with the solutions required for the peptide stimulation, the negative, and the positive controls. The plate is incubated during 15–20h to allow the specific antibodies to bind the IFN-γ. (3) Splenocytes are then washed away before the addition of the biotinylated detection antibody that binds to the previously captured IFN-γ. (4) Enzyme-conjugated streptavidin is then added to bind to the detection antibody. (5) After incubation and washing, the BCIP-NBT substrate is added, producing purple/black spots that can be quantified using appropriate analysis software (Zhang and Lehmann, Methods in Molecular Biology, 2012). Created with Smart Servier Art.

- Empty the wells as previously and wash the plate once with 100 µL of PBS 1× per well.
- Add 100 µL of blocking buffer to every well.

  For one non-sterile plate, dissolve 0.2g of dry skimmed milk in 10 mL of PBS 1× to obtain blocking buffer.

  For one sterile plate, dilute 5 mL of liquid milk in 5 mL of PBS 1×.
- Cover the plate and incubate at RT for 2h.
- Empty the wells as previously and thoroughly wash 3 times with 100 µL of PBS 1× per well.

  If you wish, you can then dry the plate and place it at 4°C for several weeks. Simply add 200 µL of PBS 1× for 1h, then continue with the next step.

3.5.2 Detection of tumor-specific CD8⁺ T cells

- Add 100 µL of cell suspension per well, i.e., 500,000 cells per well (**see** Figs. 3 and 4).
- Add 100 µL per well with a solution varying according to the well condition:
  - X-VIVO 15 (negative control).
  - X-VIVO 15 containing peptides derived from tumor- antigens. Here, a nonmutated H-2 L^d-restricted peptide antigen derived from gp70 amino acids 423–431, referred to as AH1 (SPSYVYHQF), which is the dominant target for the $CD8^+$ T cell responses against the CT26 colorectal tumor.
  - X-VIVO 15 containing PMA-ionomycin (positive control).

For each mouse, ideally prepare 1 well with a positive control, 3 wells with peptide stimulation and 3 wells as negative controls.

- Cover the plate and incubate at 37 °C in a CO_2 incubator for an appropriate length of time (15–20 h).

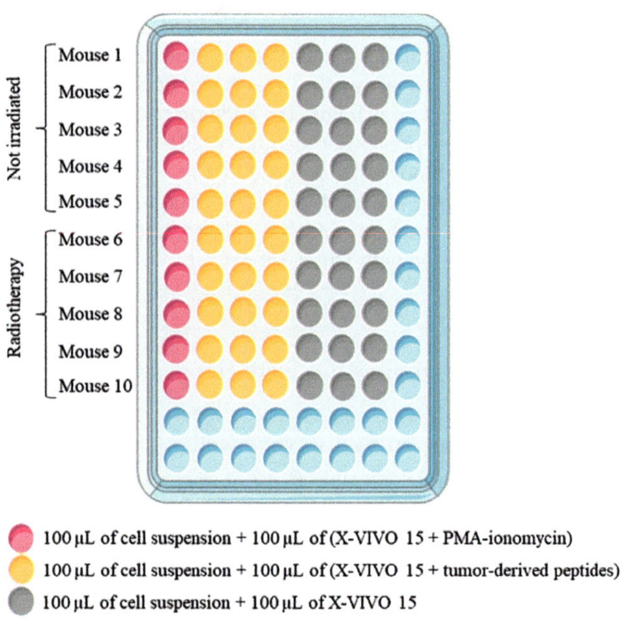

Fig. 4 Example of an IFN-γ ELISpot plate ready to incubate for 15–20 h. Experiment comparing IFN-γ secretion by splenocytes after peptide stimulation in five mice treated with radiation therapy (6–10) versus five untreated mice (1–5). Note that for each mouse (spleen), 1 positive control well, 3 negative control wells and 3 wells with tumor-peptide stimulation were used. Created with Smart Servier Art.

- Empty the wells and remove excess solution then add 100 µL of 0.05% Tween PBS to each well.

  For one plate, dilute 50 µL of Tween 20 in 100 mL of PBS 1×.
- Incubate the plate at 4 °C for 10 min.
- Empty the wells as previous and wash the plate 3 times with 100 µL of 0.05% Tween PBS.
- Add 100 µL of diluted detection antibody to each well **(see Note 10)**.
- Cover the plate and incubate at RT for 1 h 30 min.
- Empty the wells as previous and wash the plate 3 times with 100 µL of 0.05% Tween PBS.
- Add 100 µL of diluted Streptavidin-AP conjugate to every well **(see Note 10 and Note 11)**.

  For one plate, dilute 10 µL of Streptavidin-AP conjugate into 10 mL of Dilution Buffer and mix well.
- Cover the plate and incubate at RT for 1 h.
- Empty the wells and wash the plate 3 times with 100 µL of 0.05% Tween PBS.
- Peel off the plate bottom and wash both sides of the membrane 3 times under running distilled water. Once washing is complete, remove any excess solution by repeated tapping on absorbent paper.
- Add 100 µL of ready-to-use BCIP/NBT substrate to every well **(See Note 12)**.
- Incubate the plate for 5–15 min protected from light, monitoring spot formation visually throughout the incubation period to assess sufficient color development.
- Empty the wells and rinse both sides of the membrane 3 times under running distilled water. Completely remove any excess solution by gentle repeated tapping on absorbent paper.

3.6 ELISpot data analysis

- Allow the wells to dry before reading the plates. Plates should be stored at RT away from direct light. Note that color may fade over prolonged periods.
- Each spot reflects a single cell's secretory activity (Fig. 5).
- The frequency of the spots corresponding to the cytokine producing cells can be determined using an appropriate ELISpot reader and analysis software.
- For automated ELISpot data analysis, the Cellular Technology Limited (CTL) ImmunoSpot® software can be used as described in: "Objective,

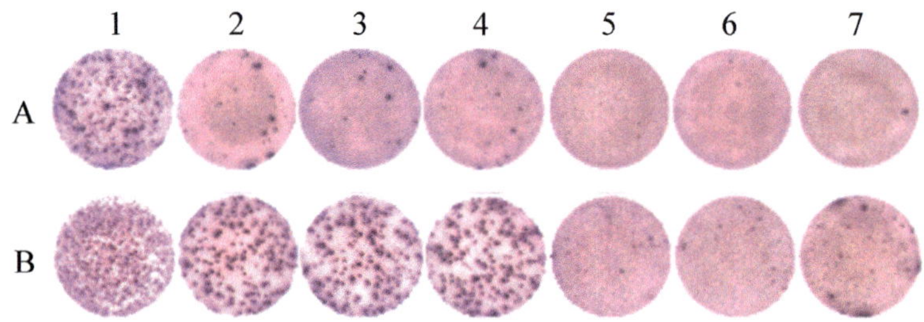

Fig. 5 A representative example of an IFN-γ ELISpot assay result. A wells correspond to an untreated mouse; B wells correspond to an irradiated mouse. A1 and B1 are the positive control wells (deep purple color with confluent cells and no white background). A5, A6, A7 and B5, B6, B7 are the negative control wells (no or very few spots as expected). A2, A3, A4 and B2, B3, B4 represent the wells where splenocytes were incubated with tumor-specific derived peptides. Contrary to the untreated mouse, the irradiated mouse displayed intense and frequent purple spots that correspond to T cells secreting IFN-γ.

user-independent ELISPOT data analysis based on scientifically validated principles." Methods Mol Biol. 2012 (Zhang & Lehmann, 2012).

- The number of specific T cells expressed as spot-forming cells per 10^5 cells is calculated after subtracting negative control values (background). A response is considered positive if the number of spots is >10 spots/10^5 cells and >2-fold background (Moodie et al., 2010).

4. Notes

(1) Disposables catalog numbers are provided as a reference; however, equivalent products may be purchased from other providers.

(2) Mice are maintained under specific-pathogen-free conditions with daily cycles of light and darkness (12h/12h), in line with ethical guidelines. No manipulation is performed during the first seven days after arrival, to allow animals to acclimatize to their new environment. For obtaining the representative images, all experimental studies were approved by the local ethics committee in accordance with the European Union's Directive 2010/63.

(3) You can change the culture medium 6h after thawing, as thawing causes many cells to die. This can limit the proliferation of viable cells.

(4) You can change the culture medium 6h after thawing, as thawing causes many cells to die. This can limit the proliferation of viable cells.

(5) Cells used for engraftment should have a viability of 90% or more.

(6) The irradiation procedure has been validated for TV ranging from 60 to 400 mm^3.

(7) Only non-filterable whitish fibrinous material remains above the filter.

(8) This intermediate dilution with PBS 1× will help you distinguish viable cells from dead cells. If counting is made difficult by the presence of excess red blood cells, you can repeat the lysis step.

(9) Check that the color of the membrane at the bottom of the wells changes from white to grayish («wet look»), indicating rehydration. If not, repeat. The most common error is the incorrect dilution of ethanol. You can extend the contact time between the ethanol solution and the membrane by tens of seconds if you see that the membranes are not completely rehydrated.

(10) For optimal performance, prepare the reconstituted antibody dilution and Streptavidin-AP dilution immediately prior to use.

(11) Do not keep this solution for further experiments. To avoid non-specific background, it is recommended to filter the working solution using a disposable syringe and a 0.2 μm filter disc.

(12) Filter the solution using a syringe and a 0.2 μm filter disc if precipitates occur.

5. Conclusion

The ELISpot assay allows the detection of low-frequency antigen-specific T cells that secrete cytokines and effector molecules. Therefore, cytotoxic T cell monitoring by ELISpot is a gold standard for the evaluation of antigen-specific T cell immunity. The good sensitivity and reproducibility of this method resulted in a wide range of applications in research, diagnostic field, and therapeutics as ELISpot may be a useful to predict clinical benefit after therapeutic immune modulation.

Acknowledgments

The authors would like to thank the French "Ligue contre le cancer", the Bourgogne Franche Comté regional council, the ARC, and the Cancéropôle EST for funding their work on this topic. We would like to thank Pr Adotévi for permitting us to use the animal facility and associated laboratory. Figures 3 and 4 were partly generated using Servier Medical Art, provided by Servier, licensed under a Creative Commons Attribution 3.0 unported license.

Declaration of interests

The authors declare no conflict of interest.

Ethical statement

Mice were housed in our animal facility. All experimental studies were approved by the small animal ethics committee and carried out in accordance with the European Union's Directive 2010/63.

References

Ahrends, T., Busselaar, J., Severson, T. M., Bąbała, N., de Vries, E., Bovens, A., et al. (2019). CD4+ T cell help creates memory CD8+ T cells with innate and help-independent recall capacities. *Nature Communications, 10*, 5531.

Borst, J., Ahrends, T., Bąbała, N., Melief, C. J. M., & Kastenmüller, W. (2018). CD4+ T cell help in cancer immunology and immunotherapy. *Nature Review Immunology, 18*, 635–647.

Hensel, J. A., Khattar, V., Ashton, R., & Ponnazhagan, S. (2019). Characterization of immune cell subtypes in three commonly used mouse strains reveals gender and strain-specific variations. *Laboratory Investigation, 99*, 93–106.

Laheurte, C., Dosset, M., Vernerey, D., Boullerot, L., Gaugler, B., Gravelin, E., et al. (2019). Distinct prognostic value of circulating anti-telomerase CD4+ Th1 immunity and exhausted PD-1+/TIM-3+ T cells in lung cancer. *British Journal of Cancer, 121*, 405–416.

Masopust, D., Sivula, C. P., & Jameson, S. C. (2017). Of mice, dirty mice, and men: Using mice to understand human immunology. *Journal of Immunology, 199*, 383–388.

Mirjolet, C., Ladjohounlou, R., Baude, J., & Bessières, I. (2023). Novel platform for subcutaneous tumor irradiation in mice. *Methods in Cell Biology*. Elsevier.

Möbs, C., & Schmidt, T. (2016). Research techniques made simple: Monitoring of T-cell subsets using the ELISPOT assay. *Journal of Investigative Dermatology, 136*, e55–e59.

Moodie, Z., Price, L., Gouttefangeas, C., Mander, A., Janetzki, S., Löwer, M., et al. (2010). Response definition criteria for ELISPOT assays revisited. *Cancer Immunology Immunothery, 59*, 1489–1501.

Zhang, W., & Lehmann, P. V. (2012). Objective, user-independent ELISPOT data analysis based on scientifically validated principles. In A. E. Kalyuzhny (Ed.), *Handbook of ELISPOT, methods in molecular biology* (pp. 155–171). Totowa, NJ: Humana Press.

CHAPTER FOUR

Tumor slice culture system for *ex vivo* immunotherapy studies

Leire Arrizabalaga[a,b], Joan Salvador Russo-Cabrera[a,b], Virginia Belsúe[a,b], Pedro Berraondo[a,b,c,†], Ignacio Melero[a,b,c,d,e,†], and Fernando Aranda[a,b,*,†]

[a]Program of Immunology and Immunotherapy, Cima Universidad de Navarra, Pamplona, Spain
[b]Navarra Institute for Health Research (IDISNA), Pamplona, Spain
[c]Centro de Investigación Biomédica en Red de Cáncer (CIBERONC), Madrid, Spain
[d]Department of Oncology and Department of Immunology and Immunotherapy, Clínica Universidad de Navarra, Pamplona, Spain
[e]Nuffield Department of Medicine and Oxford Center for Immuno-Oncology, University of Oxford, Oxford, United Kingdom
*Corresponding author: e-mail address: faranda@unav.es

Contents

† Share senior co-authorship.

Methods in Cell Biology, Volume 189
ISSN 0091-679X
https://doi.org/10.1016/bs.mcb.2024.05.007

Abstract

Personalized immunotherapy is emerging as a promising approach for cancer treatment, aiming to harness the patient's own immune system to target and eliminate tumor cells. One key aspect of developing effective personalized immunotherapies is the utilization of tumor slices derived from individual patient tumors. Tumor slice models retain the complexity and heterogeneity of the original tumor microenvironment, including interactions with immune cells, stromal elements, and vasculature. These *ex vivo* models serve as valuable tools for studying tumor-immune interactions and for testing the efficacy of immunotherapeutic agents tailored to the specific characteristics of each patient's tumor. In this chapter, we set up a protocol for immunotherapy strategies in mouse models highlighting their translational potential to guide treatment decisions and improve therapeutic outcomes in cancer patients.

1. Introduction

1.1 *Ex vivo* systems to personalized cancer immunotherapy

Immunotherapy has revolutionized cancer therapy and offers promising opportunities with the potential to induce sustained remissions in patients with neoplastic disease. Based on efficacy, active immunotherapies aimed at inducing or augmenting anti-tumor immune responses are routinely used in clinical practice since recent years (Melero et al., 2015; Whiteside, Demaria, Rodriguez-Ruiz, Zarour, & Melero, 2016). Monoclonal antibodies (mAbs) that block the co-inhibitory PD-1/PD-L1 receptor-ligand pair are the best-known clinically developed examples. In the field of immunotherapy, biomarkers to identify patients who would benefit from therapy are yet very imprecise and, hence, a matter of intensive study (Zou, Wolchok, & Chen, 2016). In this era of personalized medicine, the goal is to predict the most appropriate treatment for the individual patient. Only 15–40% of patients respond to ICI-based immunotherapy and these agents can sometimes trigger severe auto-immune reactions (Sharma, Hu-Lieskovan, Wargo, & Ribas, 2017). Thus, it is critical for the studies of personalized immunotherapy to save treatment costs, spare patients from side effects, and increase the immunotherapy efficacy. For this reason, it is an urgent need to develop reproducible and informative screening platform systems to identify, optimize, and clinically orient the best immunotherapy strategy (single agents or combinations).

In this regard, two key aspects in the tumor context must be considered when developing models to study the effect of immunotherapeutic agents: *tumor heterogeneity, tumor immune microenvironment* (Jimenez-Sanchez et al., 2017). Selecting a suitable model system is essential for the right purpose.

Some of these studies/models are based on preclinical *in vitro* and *in vivo* animal models that poorly mimic the complexity of each cancer patient. Alternatively, excision assays involving the removal of a tumor fragment, bone marrow, and immune tissues from the host before or after treatment to determine the effects of immunotherapy in *ex vivo* assays are considered valuable preclinical tools. Various platforms have been applied to generate primary cultures from individual tumors with different tumor complexity and heterogeneity (Fig. 1), including 2D monolayer culture of dissociated tumor cells; 3D organoids cultures; Patient-derived mouse xenograft (PDX) cultures; and Organotypic/Tumor slice cultures (Meijer, Naipal, Jager, & van Gent, 2017).

These methods are to be developed through the coordinated collaboration of clinicians and basic researchers and summarized in the sentence "*From bedside to bench and back again*" to speak of translational and reverse translational research.

If patients donate cancer tissue from their surgery upon informed consent. Then, tumor tissues can be tested in the different *ex vivo* models mentioned above upon exposure to therapeutic agents. Each *ex vivo* model system has its advantages and pitfalls/limitations (Fig. 2): (*i*) **2D cultures from primary culture** approaches provide high-throughput capability within a simplified cellular context. However, these systems are limited because they lack the native tumor microenvironment and spatial organization; (*ii*) **3D organoid cultures** are synthetic tissue systems that incorporate basic features of tissue architecture. Organoids have the advantage of including cells within a more physiological and morphologically complex context. However, these systems have a lower-throughput capability than cell culture systems and take weeks to months to generate and do not contain immune cells that need to be added artificially to them (Kapalczynska et al., 2018); (*iii*) **PDX models** have already been used in a diverse range of preclinical cancer immunotherapy research studies (Sanmamed, Chester, Melero, & Kohrt, 2016). Although PDXs are excellent *ex vivo* platforms for precision medicine, they have several limitations. The time of generation (can take as long as 6 months, or longer, to establish) or the low tumor engraftment rate; (*iv*) **Organotypic tumor slice cultures** provide an opportunity to study

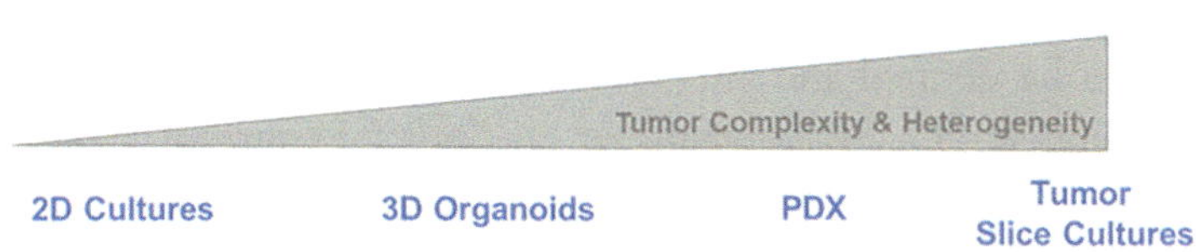

Fig. 1 Tumor complexity and heterogeneity in the different *ex vivo* systems.

	Monolayer Cultures	Organoid Cultures	PDX	Tumor Slice Cultures
Preservation of Tumor Microenvironment	-	+	++	+++
Resemblance to *in vivo* response	-	+	+++	++
Model system generation	+++	+	-	++
Short generation time	+++	+	-	++
Extended *ex vivo* cell viability	+++	+	++	-
Physiological relevance	-	+	+++	++
Costs	+++	+	-	++

Fig. 2 Classification of the more advantageous (+) to more disadvantageous (−) *ex vivo* systems.

tumor tissue behavior in the context of a relatively intact microenvironment and with a native extracellular matrix disposition, critical to test immunotherapy strategies as the most realistic microenvironment as possible. This model has potential to be exploited to personalize cancer immunotherapies (Sivakumar et al., 2019).

1.2 Tumor slice culture system

From the experimental point of view, a way to preserve tumor tissue architecture with little or no manipulation is through the culture of slices cut from fresh tumor excisions. *Ex vivo* tumor slices constitute a promising approach, as they preserve the original tumor microenvironment. These cultures can be, considered as "microtumors" that contain practically all cellular components from the tumor which they are derived from (Davies et al., 2015). With this experimental system greater reproducibility is achieved by comparative experiments performed with different slices of the same tumor explant, and a considerable number of replicate tumor slices that can be cultured under controlled conditions and treated with different agents in serial sections. For its preparations, the Leica vibratome and other different vibrating microtomes can be used. In cancer patients, for the culture tumor slices, it is essential that the tumor fragment to be used should be obtained during surgery, under aseptic conditions, and selected by a trained pathologist to get a representative sample of the particular tumor. This sample must be transported to the laboratory and processed rapidly (less than 2–3 h from surgical resection to setting the cultures of the slices). The slice thickness (depending on the experimental needs varies from 250 to 400 μm), and optimal culture conditions (nutrients, rotation, air-liquid interface, atmospheric oxygen, *etc.*) need to be defined and are critical (Jiang et al., 2017; Meijer et al., 2017; Nagaraj et al., 2018; Naipal et al., 2016; Peranzoni et al., 2017; Zhang et al., 2019). Different tumors require different culture conditions, and it is not easy to devise a universal culture protocol in these studies.

The most relevant disadvantage of the tumor slice system is its viability. Generally, cultured tumor slices retain their typical morphology and viability for at least 72h but poorly beyond that time point. In some instances, this time point might not be sufficient to get conclusive results. Some groups have been successful in maintaining viable tissue sections for 7 days to test chemotherapy agents. However, for immunotherapeutic agents, short-term cultures (48–72h) are sufficient to determine the immunomodulation of the tumor immune microenvironment (Sivakumar et al., 2019).

Tumor slices have been tested over the years, originally for cytotoxicity assays of chemotherapeutic agents or radiation therapy. It has only recently been proposed to use this methodology to evaluate the efficacy of different immunotherapies. Therefore, we consider the culture of patient-derived tumor slices as a promising methodology. Its results can be complemented with other *ex vivo* approaches and have potential to be used for personalized medicine.

In order to optimize methodology in a reproducible manner we first set up the cultures using engrafted mouse tumors that can be excised and processed. In this chapter we describe the preparative methodology employed with transplantable mouse tumors.

2. Materials and methods

2.1 Cell culture and reagents

OVA-transfected EL-4 thymoma cells, E.G7-OVA, and MC38 colorectal cancer cells were purchased from American Type Culture Collection (ATCC® CRL-2113™) and were grown in RPMI containing 10% fetal bovine serum and antibiotics. All cells were maintained at 37°C in 5% CO_2- humidified atmosphere. Monoclonal antibodies: anti-mouse 4-1BB (CD137) Ab (clone 3H3) (BioXcell), purified anti-mouse CD3 Ab (clone 17A2) (Biolegend), purified anti-mouse CD28 Ab (clone 37.51) (Biolegend). TLR agonist: ODN 1668—TLR9 agonist (CpG), Poly(I:C)—TLR3 agonist, and LPS–TLR4 agonist, were obtained from InvivoGen. PMA and Ionomycin were purchased from Sigma. Murine IFN-gamma from supernatants of tumor slice cultures were measured by ELISA kit (BD Biosciences) following the manufacturer's instructions.

2.2 Animals

Female C57BL/6 mice (6–8 weeks old; Envigo), WT or OT-I (transgenic T cell receptor designed to recognize ovalbumin peptide residues 257–264

($OVA_{257–264}$) in the context of $H2K^b$), were maintained in *pathogen-free* conditions and treated according to guidelines of our institution Cima Universidad de Navarra. All animal experiments were performed after approval by the review committee of the Gobierno de Navarra. Briefly, 5×10^5 syngeneic tumor cells MC38 or EG.7-OVA were injected subcutaneously and sacrificed as soon as they reached a volume of 1200 mm^3 for the preparation of tumor slices.

2.3 Preparation of tumor slices

E.G7-OVA and MC38 tumor pieces using a biopsy punch 5 mm (Kai medical) were embedded in low-temperature gelling agarose (Promega, V2111) before mounting on the magnetic specimen holder of a Leica VT1000S vibrating blade microtome (Vibratome) using tissue adhesive glue Vetbond 3 M. Then, automated slicing was performed with slice thickness set at 270 μm, vibration amplitude at 3.0 mm, and slicing speed at 0.6 mm/s with a sapphire blade (Ted Pella, INC) set an angle of 17°. Slicing was performed under semi-sterile conditions, without the use of a flow hood, and no contaminations were detected under these conditions. Slices were cultured within 4 h after the tumor was removed from the mouse and maintained at 4 °C in HBSS (Gibco). Tumor slices were cultured in Dulbecco's Modified Eagle Medium (DMEM)/F12 Advanced supplemented with 5% Fetal Bovine Serum, 1× GlutaMAX, 0.5× Penicillin/Streptomycin, 1× Insulin-Transferrin-Selenium and 15 mM HEPES (all from Gibco), individually on 0.4 μm, 12 mm diameter inserts (Millipore, PICM01250) with previously collagen type I coated (Corning) using 24-well plates. Briefly, to coat well cell culture insert membrane we added 50 μL of collagen gel mixture (0.025 N NaOH, 3 mg/mL rat tail collagen type I in PBS buffer), and ~100 μL of culture medium on top of inserts and 300 μL of culture medium to each well of 24-well plate outside of inserts (Fig. 3). Culturing was performed at 5% CO_2 at 37 °C and at atmospheric oxygen levels with agitation at 25 rpm. The culture medium was replaced every 24–48 h.

2.4 Hematoxylin and eosin staining

Histological tumor architecture was examined by hematoxylin and eosin (H&E) staining. It was carried out by the Histology service at Cima University of Navarra.

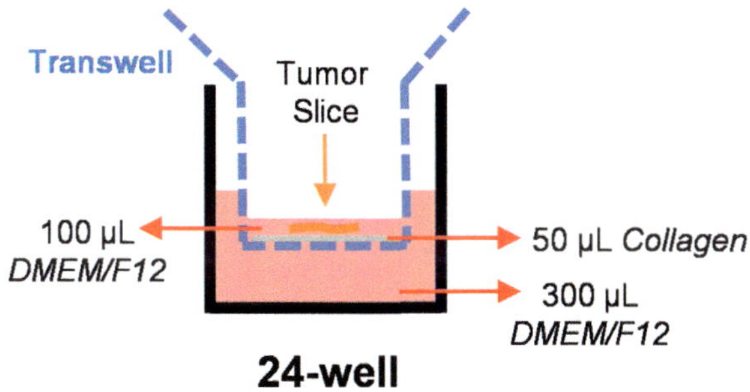

Fig. 3 Graphic scheme of tissue culture in the transwell.

2.5 AlamarBlue HS Viability Assay

The Tissue slice viability along the days was quantified with The AlamarBlue (AlaB) High-Sensitivity (HS) Cell Viability Reagent (Invitrogen, A50100). AlaB is a blue nonfluorescent dye reduced to a pink-colored, highly fluorescent resorufin by metabolically active cells. It is known that viable cells reduce the microenvironment to pinkish color. In contrast, dead or inactive cells do not change the original blue color of resazurin, the active ingredient of AlaB. During 96h of incubation with or without compounds, the explants were incubated for 4h with 10% AlaB in 200μL DMEM/F12 supplemented medium at 37°C in 24-well plate. Afterward, 100μL was collected from each sample and transferred to a 96-well microplate. Fluorescence values were read using a microplate reader (Thermo Multiskan Ascent Plate Reader) at 530nm excitation/590nm emission wavelengths.

2.6 Tumor slice dissociation

Tumor slices were dissociated using 300μL of serum-free DMEM/F12 Advanced with 100μg/mL hyaluronidase (Sigma, H3506) and 100μg/mL of Liberase TL (Roche, 5401020001) in 24-well plate for 30min. The dissociation mix was resuspended every 5min, and 10μg/mL of DNAse I from bovine pancreas (Roche, 11,284,932,001) was added during the last 10min. To stop the enzymatic digestion completed DMEM/F12 Advanced medium was added and then filtered through a 70μm cell strainer (BD Bioscience).

2.7 Flow cytometry

Surface stainings were performed by blocking cells in phosphate-buffered saline (PBS) plus 10% FBS for 30min at 4°C with rat anti-mouse

CD16/32 Fc Block (clone 2.4G2; BD Bioscience) and further tumor slices were incubated with the following labeled specific mAb: anti-mouse CD3-PE (clone 17A2; Biolegend), APC anti-mouse/human CD11b-APC (clone M1/70; Biolegend), anti-mouse CD45-FITC (clone I3/2.3; Biolegend). Expression of CD45, CD11b, and CD3 cell populations were analyzed with a FACSCanto II flow cytometer (Becton Dickinson) and FlowJo X software (Tree Star Inc.).

2.8 CFSE-labeled

Splenocytes from WT and OT-I mice were stained with 2,5 μM CFSE (Invitrogen) in PBS for 15 min at 37 °C, combined with 20 mL of 10% FBS RPMI-1640 medium at RT for splenocytes were added to the top of the EG.7-OVA tumor slices in 24-well. After 24 h co-culture of tumor slices with CFSE-labeled splenocytes were washed twice with PBS, and the slices were stained the nuclei with DAPI. Finally, slices were visualized with the Leica DMIL Inverted Trinocular Phase Contrast Integrated Modulation Contrast Fluorescence Microscope. The image process was analyzed by Fiji-ImageJ software.

3. Procedure and results

3.1 Tumor slice preparation

We optimized the tissue slice method. After testing different strategies (optimal tumor slice thickness, the impact of various culture media, tumor slice preservation, *etc.*), we set up our *ex vivo* system as described in the schematic Fig. 4 which is explained in detail in Section 2.

3.2 Assessment of viability and tumor slice structural integrity

In all tissue culture studies, viability is the most critical point for their usefulness. Different kinds of tumors might present differences in viability and require specific types of culture media or supplements. In our case, a MC38 colorectal carcinoma murine model showed good viability for at least the first 48 h (around 70%) (Fig. 5A). However, from 72 h onward, Alamar blue HS test (Fig. 5A) and the structural integrity in histology images (Fig. 5B) showed the non-viability and, in consequence, the non-appropriate use of this system with our conditions in experiments for long-term.

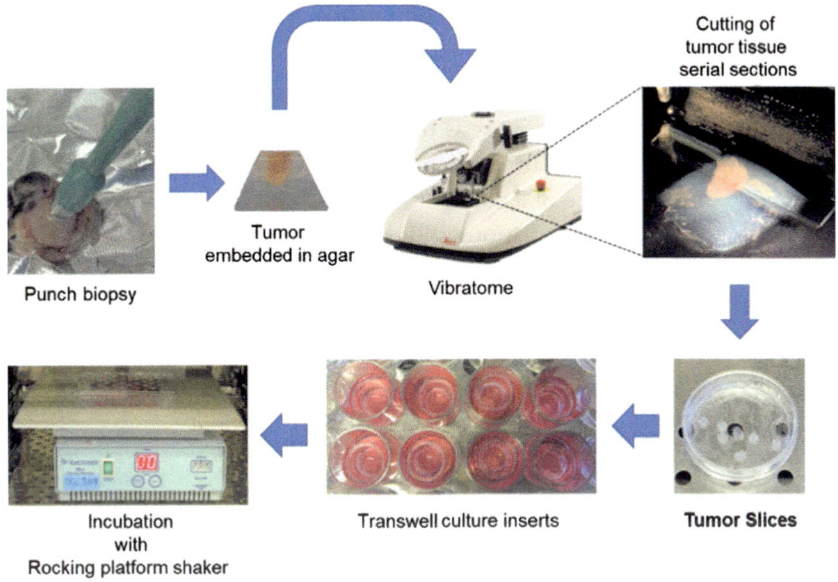

Fig. 4 Tumor slice culture workflow. Fresh tumor explants (MC38 or EG.7-OVA) were harvested from the mice on a specimen platform to embedded in agarose 4%. Tumor explant of 270 μm in thickness. Then, tumor slices were incubated in transwells with shaking at 37 °C 5% CO_2. Experiments were performed to study the viability and immune responses.

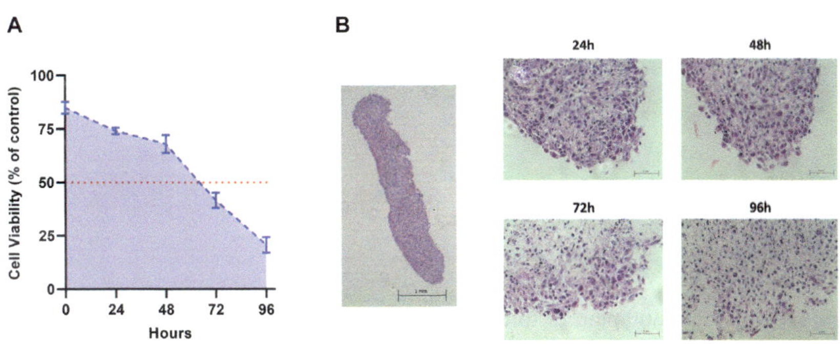

Fig. 5 Tumor slice viability and histology. (A) Tumor slices at different time points after *ex vivo* incubation (0, 24, 48, 72, and 96 h) were incubated with Alamar Blue to determine the viability ($n=3$). (B) Representative overview images of H&E-stained whole tissue slices. The scale indicates mean values.

3.3 Tumor slices depict the bulk tumor immune microenvironment

Besides keeping the tumor structure intact, one of the advantages of the tumor slice culture methodology is the representative immune infiltration when comparing the bulk, tumor slice, and serial tumor slice. We have demonstrated the resemblance of leukocytes infiltrated ($CD45^+$: lymphoid $CD3^+$ and myeloid $CD11b^+$ populations) (Fig. 6) in tumor bulk (CD45+: 49%;

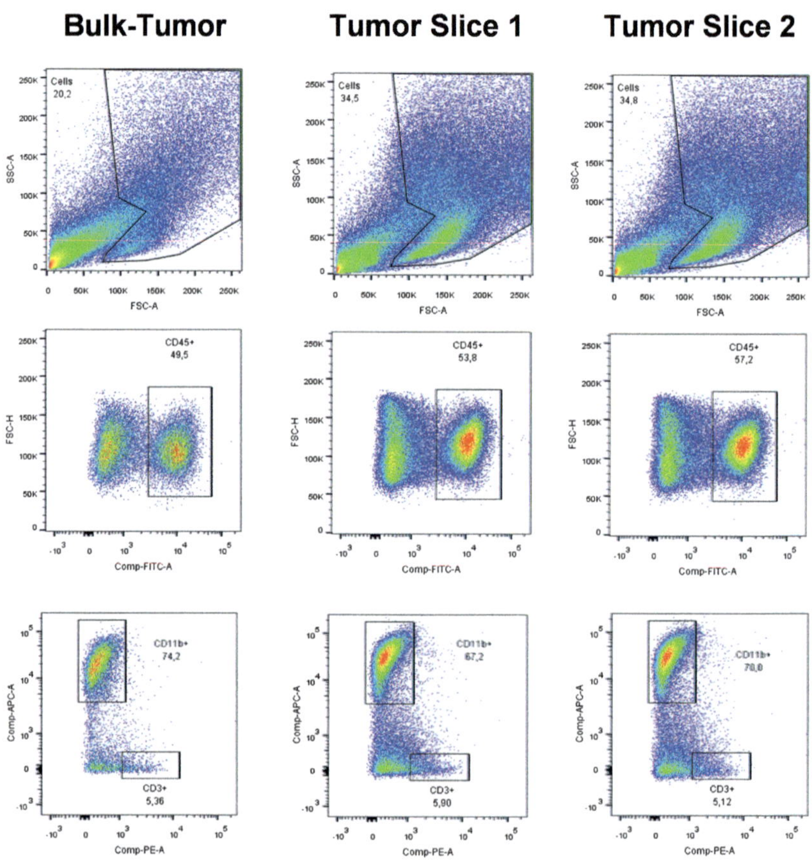

Fig. 6 Immunophenotyping of tumor slices reveals a reproducible leukocyte population between bulk-tumor, tumor slice, and serially tumor slice. The gating strategy used for identifying the immune cell composition in tumor samples is represented using polychromatic dot plots of cells from a mouse tumor-derived from subcutaneously injected syngeneic colorectal cancer cells (MC38). Cells were analyzed on the day of tumor harvest (day 0). Viable cells were gated and separated into $CD45^+$-FITC to analyze lymphocytes ($CD3^+$-PE) and myeloid ($CD11b^+$-APC). Results are representative of 2 independent experiments.

$CD3^+/CD45^+$: 5.36%; $CD11b^+/CD45^+$: 74.2%), tumor slice ($CD45^+$: 53.8%; $CD3^+/CD45^+$: 5.90%; $CD11b^+/CD45^+$: 67.2%), and the consecutive tumor slice (Tumor Slice 2) ($CD45^+$: 57.2%; $CD3^+/CD45^+$: 5.12%; $CD11b^+/CD45^+$: 70.0%) samples.

3.4 Tumor slices with immunostimulatory agents

Tumor tissue slices were serially incubated with the different immunostimulating agents. On the one side, tissue slices were stimulated with the TLR agonists, and the supernatants were collected 24 h after incubation (Fig. 7A). In parallel, T-cell immunostimulant monoclonal antibodies (mAb) were incubated for 48 h before collecting supernatants. The results showed detectable IFN-gamma levels by ELISA with TLR agonists (poly(I:C), CpG and LPS) and non-specific stimulation PMA/Ionomycin. However, when tissue slices were co-incubated with a T-Cell stimulation

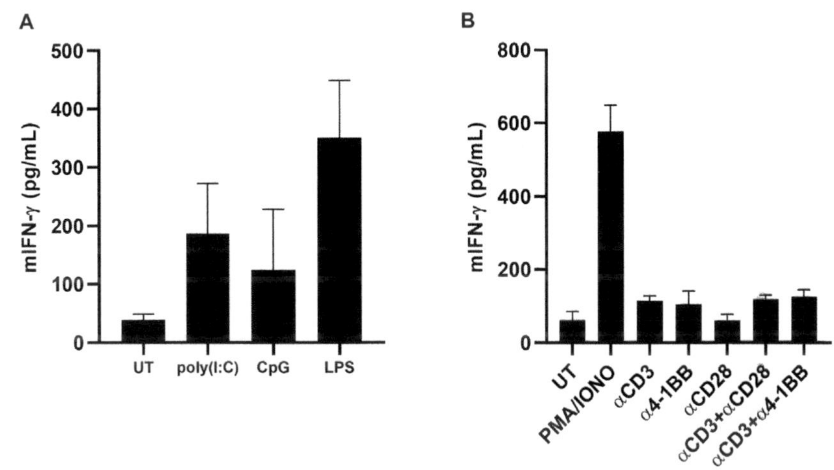

Fig. 7 IFN-gamma tumor slice production after immunostimulatory agents' incubation. Tumor slices were stimulated (from two different mice tumor tissues and using serial tumor slices to all the treatment groups) with (A) TLR agonist ligand: Poly (I:C) (10 μg/mL), CpG (5 μg/mL), LPS (2 μg/mL) during 24 h. (B) Specific monoclonal antibodies: anti-CD3 (1 μg/mL), anti-4-1BB (10 μg/mL), anti-CD28 (5 μg/mL), and non-specific PMA (20 ng/mL) plus Ionomycin (1 μg/mL) during 48 h. Then, supernatants without diluting samples were measured IFN-gamma levels by ELISA. Results are representative of 2 independent experiments.

through TCR receptor (signal 1: CD3) and co-stimulation (signal 2: CD28 or 4-1BB) with agonistic mAbs, IFN-gamma levels were not detected in comparison with the basal group by ELISA.

3.5 Specific anti-tumor CD8 response in tumor slice cultures

In general, tumors have a deficient number of infiltrating effector lymphocytes. There is a handicap to studying immunostimulants destined for effector T lymphocytes that should exert their antitumor effect directly with tumor slices. An alternative strategy could be the co-incubation of the tumor tissues with autologous T cells, pretreated with the different immunotherapy agents. To study T cell-specific behavior in the tumor tissue system, we used splenocytes from a WT or OT-I mouse with tissue slices from E.G7-OVA tumors. After 24 h, IFN-gamma levels were found in the supernatants from OT-I group (Fig. 8A), apparently more killing (less tissue edge integrity) and infiltration (intensity of CFSE) into the tissues by optical microscope (Fig. 8B).

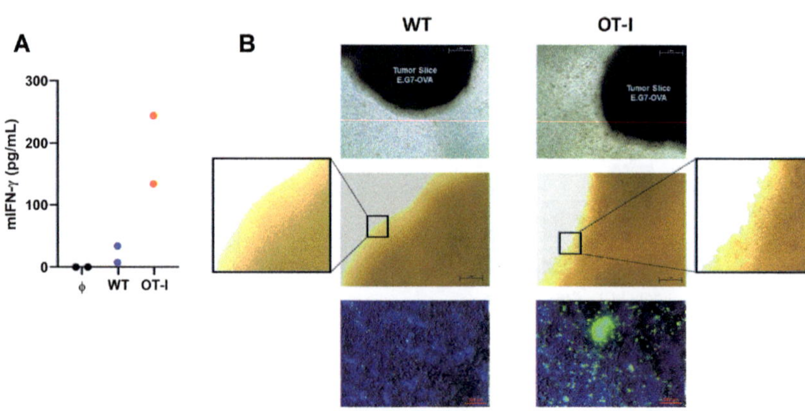

Fig. 8 Specific CD8 responses in tumor explants. EG.7-OVA tumor sections were incubated with WT, OT-I splenocytes (1×10^6), or basal group without cells (Φ). At 24 h, (A) supernatants were collected, and IFN-gamma levels were measured by ELISA. On the other hand, (B) the CFSE labeled splenocytes (2.5 μM) with DAPI tumor slices were observed by fluorescence microscopy after washed the tumor slices. Samples were obtained from two independent tumor tissues (mice) and using serial tumor slices to untreated, WT, and OT-I group.

4. Concluding remarks

Once immunotherapy has been demonstrated in multiple clinical studies that can improve the survival of cancer patients, the current challenge is to individually predict outcome prior to treatment. To achieve this goal, *ex vivo* techniques are a valuable tool to standardize personalized cancer medicine.

The opportunity to test different agents with serial cuts that replicate the tumor microenvironment makes the methodology suitable for experimentation with immunotherapy agents. In general, these tumor cultures are cut in slices of around 250-400 μm thickness with a vibratome, although there are also approaches Matrigel-embedded tumor explains set up in cultures as mini-cubes of 1 mm^3.

The use of tumor slice cultures to test immunotherapeutic agents is yet to be broadly implemented. We have proposed to study the methodology in mouse models to determine if the slice system is viable for the purposes reliably testing immunotherapeutic agents. Tissue viability is a critical factor, and we have kept these cultures viable for up to 1 week, we have reached the consensus to maintain the tumor tissue for up to 48 h. Other authors have concluded that this time range (48–72 h) should be an optimal interval to analyze immunotherapy agents in tumor slice cultures.

When tumor slices (from MC38-derived tumors) were stimulated with different immunostimulatory agents, we have seen interesting results. As shown, when the slices were incubated with TLR agonists, IFN-gamma production was detected in the supernatants at 24 h by ELISA. On the contrary, in serial sections incubated with agonistic antibodies against T lymphocytes (anti-CD3, and co-stimulation with anti-CD28 or anti-4-1BB) IFN-gamma levels were undetectable by ELISA with our protocol. Probably the reasons underlying these results are the scarcity of tumor-infiltrating lymphocytes (TILs) (around 5% $CD45^+CD3^+$) and the volume of the culture medium per section, which had to be at least 400 μL and may result in the excessive dilution of released mediators. By contrast, tumor slices can be useful to test artificial addition of tumor-specific lymphocytes, CAR-T, or to study lymphocytes previously treated with the immunostimulatory agents before co-culturing them with the tumor tissue slices. In this regard, relatively fast results (24–48 h) could be obtained (measuring activation, effector activity, *etc.*) with different laboratory *state-of-the-art* techniques, including single-cell RNA sequencing.

Ongoing and future research will define the applicability of the technology to surgical human tumor samples and with the goal of testing immunotherapy agents and their combinations, learning the feature of each case that would be ultimately associated with clinical efficacy.

Acknowledgments

This study was supported by Instituto de Salud Carlos III (PI22/00147, and PI23/00203) co-financed by Fondos Feder and Gobierno de Navarra Proyecto ARNMUNE Ref: 0011–1411-2023. P.B. receives a 2022 Leonardo Grant for Researchers and Cultural Creators (BBVA Foundation). F.A. receives a Miguel Servet I (CP19/00114) contract from ISCIII (Instituto de Salud Carlos III) co-financed by FSE (Fondo Social Europeo).

References

Davies, E. J., Dong, M., Gutekunst, M., Narhi, K., van Zoggel, H. J., Blom, S., et al. (2015). Capturing complex tumour biology in vitro: Histological and molecular characterisation of precision cut slices. *Scientific Reports*, *5*, 17187.

Jiang, X., Seo, Y. D., Chang, J. H., Coveler, A., Nigjeh, E. N., Pan, S., et al. (2017). Long-lived pancreatic ductal adenocarcinoma slice cultures enable precise study of the immune microenvironment. *Oncoimmunology*, *6*(7), e1333210.

Jimenez-Sanchez, A., Memon, D., Pourpe, S., Veeraraghavan, H., Li, Y., Vargas, H. A., et al. (2017). Heterogeneous tumor-immune microenvironments among differentially growing metastases in an ovarian cancer patient. *Cell*, *170*(5), 927–938. e20.

Kapalczynska, M., Kolenda, T., Przybyla, W., Zajaczkowska, M., Teresiak, A., Filas, V., et al. (2018). 2D and 3D cell cultures—A comparison of different types of cancer cell cultures. *Archives of Medical Science*, *14*(4), 910–919.

Meijer, T. G., Naipal, K. A., Jager, A., & van Gent, D. C. (2017). Ex vivo tumor culture systems for functional drug testing and therapy response prediction. *Future Science OA*, *3*(2), FSO190.

Melero, I., Berman, D. M., Aznar, M. A., Korman, A. J., Perez Gracia, J. L., & Haanen, J. (2015). Evolving synergistic combinations of targeted immunotherapies to combat cancer. *Nature Reviews. Cancer*, *15*(8), 457–472.

Nagaraj, A. S., Bao, J., Hemmes, A., Machado, M., Narhi, K., & Verschuren, E. W. (2018). Establishment and analysis of tumor slice explants as a prerequisite for diagnostic testing. *Journal of Visualized Experiments*, *141*.

Naipal, K. A., Verkaik, N. S., Sanchez, H., van Deurzen, C. H., den Bakker, M. A., Hoeijmakers, J. H., et al. (2016). Tumor slice culture system to assess drug response of primary breast cancer. *BMC Cancer*, *16*, 78.

Peranzoni, E., Bougherara, H., Barrin, S., Mansuet-Lupo, A., Alifano, M., Damotte, D., et al. (2017). Ex vivo imaging of resident CD8 T lymphocytes in human lung tumor slices using confocal microscopy. *Journal of Visualized Experiments*, *130*.

Sanmamed, M. F., Chester, C., Melero, I., & Kohrt, H. (2016). Defining the optimal murine models to investigate immune checkpoint blockers and their combination with other immunotherapies. *Annals of Oncology*, *27*(7), 1190–1198.

Sharma, P., Hu-Lieskovan, S., Wargo, J. A., & Ribas, A. (2017). Primary, adaptive, and acquired resistance to cancer immunotherapy. *Cell*, *168*(4), 707–723.

Sivakumar, R., Chan, M., Shin, J. S., Nishida-Aoki, N., Kenerson, H. L., Elemento, O., et al. (2019). Organotypic tumor slice cultures provide a versatile platform for immuno-oncology and drug discovery. *Oncoimmunology*, *8*(12), e1670019.

Whiteside, T. L., Demaria, S., Rodriguez-Ruiz, M. E., Zarour, H. M., & Melero, I. (2016). Emerging opportunities and challenges in Cancer immunotherapy. *Clinical Cancer Research*, *22*(8), 1845–1855.

Zhang, W., van Weerden, W. M., de Ridder, C. M. A., Erkens-Schulze, S., Schonfeld, E., Meijer, T. G., et al. (2019). Ex vivo treatment of prostate tumor tissue recapitulates in vivo therapy response. *The Prostate*, *79*(4), 390–402.

Zou, W., Wolchok, J. D., & Chen, L. (2016). PD-L1 (B7-H1) and PD-1 pathway blockade for cancer therapy: Mechanisms, response biomarkers, and combinations. *Science Translational Medicine*, *8*(328). 328rv4.

CHAPTER FIVE

Identification and analysis of alloreactive T lymphocytes from peripheral blood mononuclear cells

Alberto Susana[a], Giovanni Galletti[b], Gabriele De Simone[a], Chiara Camisaschi[a], and Enrico Lugli[a,*]

[a]IRCCS Humanitas Research Hospital, Rozzano, Milan, Italy
[b]Department of Molecular Biology, School of Biological Sciences, University of California, San Diego, San Diego, CA, United States
*Corresponding author: e-mail address: enrico.lugli@humanitasresearch.it

Contents

Abstract

Alloreactive T-cell responses against mismatched MHC or minor histocompatibility antigens may result in deleterious graft-versus-host disease (GVHD) and increased morbidity and mortality in allogeneic hematopoietic stem cell transplantation (allo-HSCT). Nevertheless, these T-cell responses may be directed against residual tumor cells (the graft-versus-tumor effect, GVT), thus preventing relapse of the disease. Recent findings have shown that $CD45RA^+$ naïve T cells, but not $CD45RA^-$ memory T cells are the major contributors to GVHD, thus leading to clinical trials where $CD45RA^+$-depleted, memory-enriched T-cell products are adoptively transferred following allo-HSCT to prevent GVHD and enhance immune reconstitution. However, residual alloreactivity may still be present in the memory T-cell compartment, thus contributing to prevent

Methods in Cell Biology, Volume 189
ISSN 0091-679X
https://doi.org/10.1016/bs.mcb.2024.05.011

disease relapse by GVT. Here, we describe a simple cell-based protocol to identify alloreactive naïve and memory T cells by co-culturing T-cell subsets and third-party antigen-presenting cells. The responding cells are identified following dilution of carboxyfluorescein succinimidyl ester (CFSE) and upregulation of the activation marker CD25. These CFSE-diluting cells can be further phenotyped by high-dimensional flow cytometry, or purified with a cell sorter for downstream genomic and functional assays.

1. Introduction

Alloreactivity remains one of the major obstacles in allogeneic hematopoietic stem cells transplantation (HSCT). The major determinants of alloreactivity are thought to be mismatched major histocompatibility complex (MHC) loci between the donor and the host: upon recognition of host peptide:MHC (p:MHC) complexes expressed on the surface of all nucleated cells in the case of MHC class I, or antigen presenting cells (APC) in the case of MHC class II, donor T lymphocytes undergo activation, proliferate and respond by producing proinflammatory cytokines or cytotoxic molecules (Warren et al., 2012). It is estimated that the precursor frequency of T cells recognizing allogeneic MHC molecules is in the range of 1–10% of the total T-cell repertoire (Siu, Surendrakumar, Richards, & Pettigrew, 2018). In addition, peptides derived from self-proteins carrying host polymorphisms, known as minor histocompatibility antigens, can be presented on the surface of donor APCs. Although T cells targeting these minor antigens are initially in the range of 0.01–0.1%, they can nevertheless exert deleterious T-cell responses against the host and be responsible of graft failure (Eckle, Rossjohn, & McCluskey, 2013; Felix & Allen, 2007).

In hematological malignancies, donor T cells may attack healthy host tissues, with a preference for the skin, the gut and the liver, and thus cause the so-called graft versus host disease (GVHD), a major cause of morbidity and mortality following allogeneic HSCT (Ferrara & Chaudhry, 2018). However, at the same time, these alloreactive T cells can recognize mismatched MHCs on the surface of malignant cells, thereby exerting beneficial anti-tumor immunity collectively known as graft versus tumor (GVT) (Gratwohl et al., 2017). Unfortunately, GVHD and GVT share similar biological mechanisms of action, therefore it has been difficult to uncouple these two processes towards safer, more effective HSCT procedures (Fowler, 2006).

T cells are the major cellular players in GVHD/GVT and can be roughly divided in naïve and memory cells (Mahnke, Brodie, Sallusto, Roederer, & Lugli, 2013). Memory T cells, previously primed by exogenous antigens (e.g., viruses and bacteria), are clonally expanded, while naïve T cells are polyclonal, and are capable of potentially recognizing any type of foreign antigen. Both cell subsets from the same individual are obviously restricted towards the same host MHC molecules, and could ideally mediate alloreactive responses towards mismatched MHCs from the host. However, adoptive cell transfer experiments with purified populations have shown that more potent alloreactivity is exerted by naïve T cells, probably owing to their open T-cell repertoire, which enhances their probability to recognize mismatched MHCs compared to memory T cells (Foster et al., 2004; Macedo et al., 2009). These preclinical results inspired clinical trials where purified memory T cells from MHC matched or partially mismatched (haploidentical) donors are added back during or after allogeneic HSCT with the purpose of limiting GVHD while promoting immune reconstitution (Bleakley et al., 2015; Castagna et al., 2021; Shook et al., 2015; Touzot et al., 2015). The conducted trials reported minimal, or no clinical signs of GVHD in patients with blood cancers receiving the purified memory T cells. This is somewhat surprising, giving that alloreactive memory T-cell responses can be readily measured in vitro (Roberto et al., 2015). Thus, identifying the mechanisms at the basis of alloreactive memory vs. naïve T-cell responses can be beneficial to design novel adoptive cell therapy approaches to maximize GVT while limiting GVHD.

Here, we propose a protocol (Fig. 1) to expand alloreactive T cells from peripheral blood mononuclear cells (PBMC) isolated from buffy coat, by

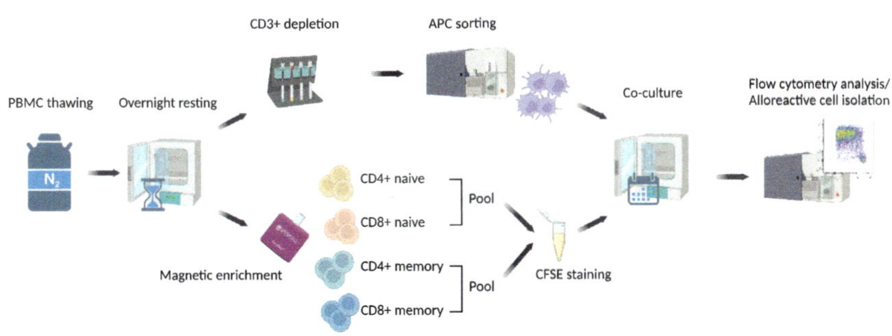

Fig. 1 Schematic overview of the procedure. PBMC: peripheral blood mononuclear cells; APC: antigen presenting cells; CFSE: carboxyfluorescein succinimidyl ester. Created with Biorender.com.

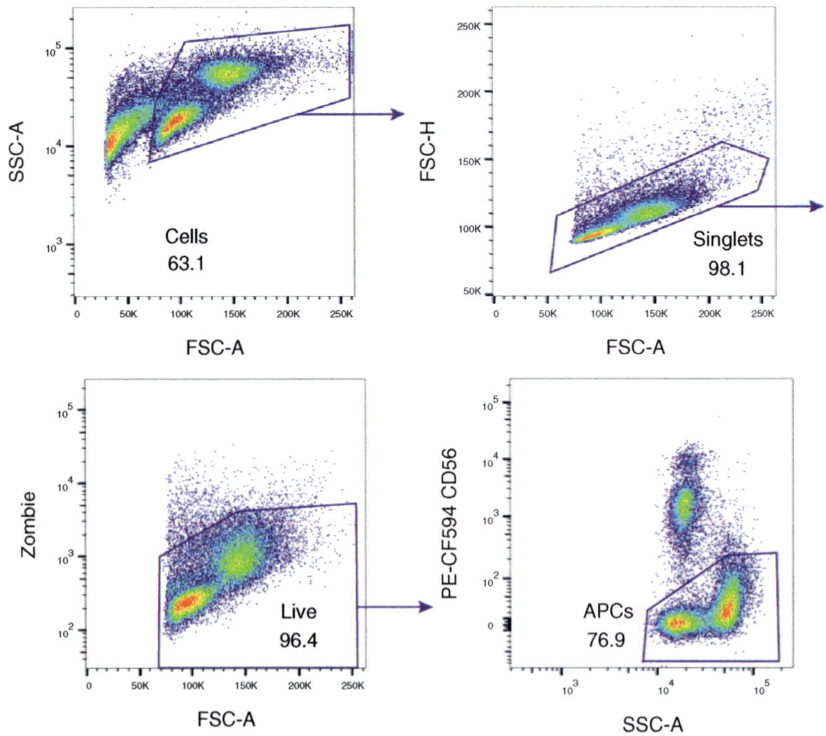

Fig. 2 Flow cytometry gating strategy to sort APCs. A first gate on side scatter-area (SSC-A) vs. forward scatter-area (FSC-A) is used to discriminate cells from debris, then singlets are isolated on the basis of FSC-H vs. FSC-A combination. Live cells are selected as negative for the Zombie viability dye and natural killer (NK) cells are excluded on the basis of CD56 expression. Numbers indicate percentages identified by the gates.

co-culturing them with allogeneic APCs obtained from third-party donors (Fig. 2). Expansion of alloreactive clones is tracked by measuring the dilution of the proliferation dye carboxyfluorescein succinimidyl ester (CFSE) and staining of the activation marker CD25 (Fig. 3A). The alloreactive T cells identified as CFSE diluted, CD25+ by flow cytometry can be further characterized by high-dimensional single cell analysis or purified with a cell sorter for downstream molecular and functional assays.

2. Materials

2.1 Common disposables

- Falcon® 50 mL high-clarity polypropylene (PP) conical centrifuge tubes (#352070, Corning) (see *Note 1*)
- Falcon® 15 mL high-clarity PP conical centrifuge tubes (#352096, Corning) (see *Note 1*)

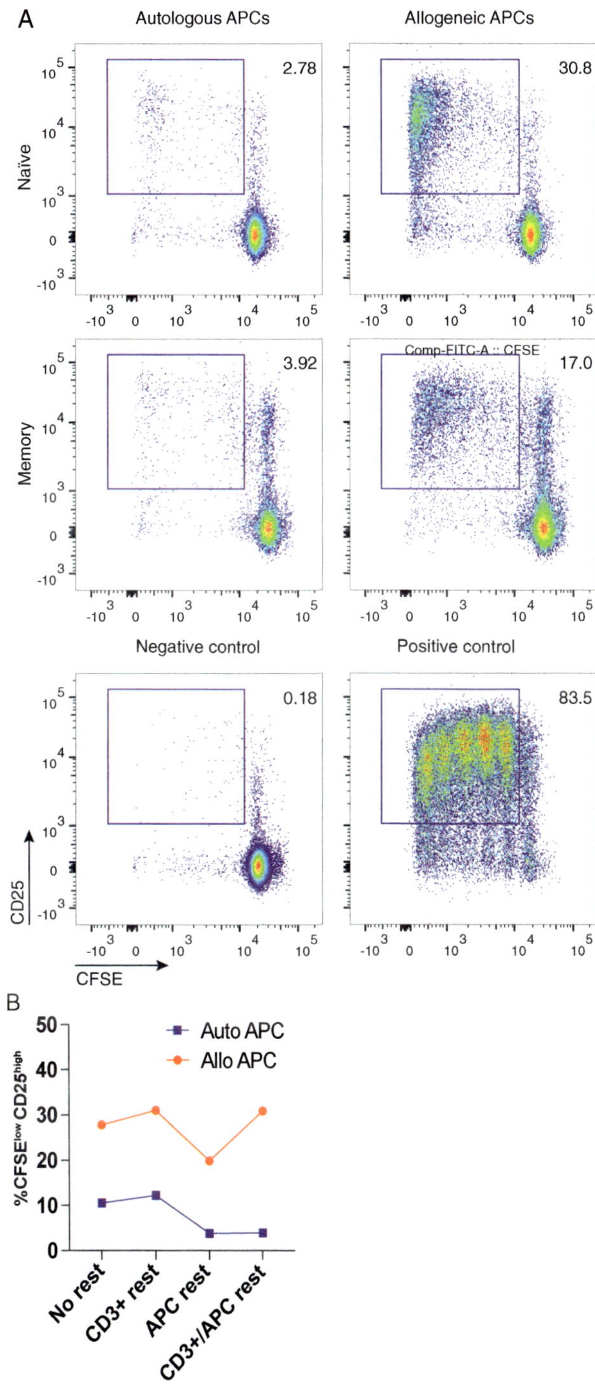

Fig. 3 See figure legend on next page.

- 1.5 mL microcentrifuge tubes (e.g., SafeSeal reaction tube, 1.5 mL, PP, #72.706, SARSTEDT) (see *Note 1*)
- FACS tubes (e.g., Falcon® Round Bottom High-clarity Polypropylene Tubes, #352008, Corning) (see *Note 1*)
- Falcon® 40 µm Cell Strainer (#352340, Corning) (see *Note 1*)
- LD Columns (#130-042-901, Miltenyi Biotec) (see *Note 1*)
- Falcon® 14 mL Round Bottom High Clarity PP Test Tube (#352059, Corning) (see *Note 1*)
- Falcon® Polystyrene Tissue Culture Plate, 96-well, U-Bottom TC-treated, with Low Evaporation Lid (#353077, Corning) (see *Note 1*)
- Falcon® 75 cm^2 Rectangular Canted Neck Cell Culture Flask with Vented Cap (#353136, Corning) (see *Note 1*)

2.2 Cells and reagents

- PBMCs from at least 2 donors, obtained under specific IRB approval (see *Note 2*)
- R10 medium: RPMI 1640 Medium without L-Glutamine with Phenol Red (#ECB9006L, EuroClone) (see *Note 1*), 1% *v*/v L-Glutamine 100× (200 mM) (#ECB3000D, EuroClone) (see *Note 1*), 1% v/v Penicillin/Streptomycin 100× (#ECB3001D, EuroClone) (see *Note 1*), 10% v/v Fetal Bovine Serum (FBS) South America origin EU Approved (#ECS0180L, EuroClone) (see *Note 1*)
- Dulbecco's Phosphate Buffered Saline (PBS) (#D8537, Merck) (see *Note 1*)
- Flow cytometry staining buffer: FBS (#ECS0180L, EuroClone) 2% v/v in PBS (see *Note 1*)
- PE-Cyanine7 mouse anti-human CD3 (#300420, BioLegend®) (see *Note 3*)
- PE-CF594 mouse anti-human CD56 (NCAM-1) (#562289, BD Horizon™) (see *Note 3*)
- PE mouse anti-human CD4 (#344606, BioLegend®) (see *Note 3*)

Fig. 3 Identification of alloreactive naïve and memory T cells. (A) CFSE dilution and CD25 expression by naïve and memory T cells following co-culture with autologous and allogeneic APCs. Negative and positive controls are T cells in IL-2 alone or stimulated with anti-CD3/CD28 beads + IL2 (20 IU/mL), respectively. The gate identifies cells diluting CFSE and expressing the activation marker CD25 ($CFSE^{low}$ $CD25^{high}$). Numbers indicate percentages identified by the gate. (B) Role of $CD3^+$ and APC resting on the fraction of activated $CD3^+$ T cells following incubation with allogeneic (red line) or autologous APC (blue line). Resting both $CD3^+$ and APCs decreases background activation in response to autologous APCs. APC: antigen-presenting cell; CFSE: carboxyfluorescein succinimidyl ester.

- BV786 mouse anti-human CD8 (#563823, BD Horizon™) (see *Note 3*)
- BV650 mouse anti-human CD25 (#563719, BD Horizon™) (see *Note 3*)
- Zombie Aqua™ Fixable Viability Kit (#423102, BioLegend®) (see *Note 1*)
- Anti-PE MicroBeads (#130-048-801, Miltenyi Biotec) (see *Notes 1, 4*)
- autoMACS® Rinsing Solution (#130-091-222, Miltenyi Biotec) (see *Note 1*)
- MACS® BSA Stock Solution (#130-091-376, Miltenyi Biotec) (see *Note 1*)
- RoboSep™ Buffer (#20104, STEMCELL™ Technologies) (see *Note 1*)
- Dynabeads™ Human T-Activator CD3/CD28 for T-Cell Expansion and Activation (#11131D, Thermo Fisher Scientific) (see *Note 1*)
- Human IL-2 IS, research grade (#130-097-743, Miltenyi Biotec) (see *Note 1*)

2.3 Equipment

- Laboratory biosafety cabinet (Class II) (e.g., Purifier Logic+ Class II Type A2 Biological Safety Cabinet #320419101, Labconco) (see *Note 1*)
- Benchtop centrifuge (e.g., Centrifuge 5810/5810R, Eppendorf) (see *Note 1*)
- High-Speed centrifuge (e.g., Centrifuge 5430/5430R, Eppendorf) (see *Note 1*)
- Vortex shaker (e.g., VORTEX 3, IKA) (see *Note 1*)
- Cell sorter (e.g., FACSAria™ III Cell Sorter, BD) (see *Note 1*)
- Magnet for cell separation (e.g., MojoSort™ Magnet, #480019, BioLegend®) (see *Note 1*)
- Humidified cell culture incubator (e.g., CO_2 incubator MCO-18AC-PE, #390-0588P, Panasonic) (see *Note 1*)
- Benchtop cell counter (e.g., Countess™ 3 Automated Cell Counter, #AMQAX2000, Invitrogen™) (see *Note 1*)

2.4 Software

- BD FACSDiva™ (BD) (see *Note 5*)
- FlowJo v. 10.10 (BD) (see *Note 6*)

3. Methods

1. PBMCs are isolated from the peripheral blood of at least 2 donors by Ficoll centrifugation, according to standard procedures, and are used fresh or

following cryopreservation in FBS + 10% DMSO (see *Note 7*). The amount of blood to be collected is dependent on the specific applications (see *Note 8*). Steps described below to isolate both APCs and naïve and memory T cells are to be performed for the two donors in parallel.

2. Cryopreserved PBMCs are thawed, resuspended in pre-warmed R10 medium and counted. R10 medium is prepared with 10% FBS, 1× Penicillin-Streptomycin-Glutamine (corresponding to 100 U/mL penicillin, 100 μg/mL streptomycin, 0.3 mg/mL glutamine) (see *Note 9*).
3. Cells are resuspended in R10 at a density of 10^7 cells/mL and between 15 and 50 mL are incubated in T75 flasks at 37 °C overnight (approximately 18 h). This resting step is recommended to lower the percentage of cells proliferating in response to autologous APCs (Fig. 3B) and to reduce cell mortality (see *Note 10*).
4. The day after, cells are collected in a 50 mL tube, washed twice with PBS−/− by centrifuging at 500 × *g* for 10 min and resuspend in 30 mL PBS−/−.
5. Cell suspensions are split in two equal fractions: fraction 1 is used to isolate APCs with a cell sorter (steps 6–11), fraction 2 to separate naïve and memory T cells by magnetic enrichment (steps 12–13). Perform these procedures in parallel.
6. To allow a faster and cleaner separation of APCs by FACS-based sorting, proceed with magnetic depletion of $CD3^+$ T cells from fraction 1.
7. Label cell suspension with the pretitrated PE-conjugated anti-CD3 antibody, as described (Lugli, Zanon, Mavilio, & Roberto, 2017) (see *Note 11, 12, 13*). Anti-CD3 conjugated to PE tandem dyes (such as PE-Cy5 and PE-Cy7) can also be used, although preliminary testing is advised (see *Note 14*).
8. Magnetically deplete PE-positive cells with anti-PE microbeads following manufacturer instructions (Miltenyi Biotec, catalog. n. 130-048-801).
9. Collect the untouched $CD3^-$ fraction, wash with PBS by centrifuging at 500 × *g* for 10 min and resuspend in 250 μL PBS−/− + 2% FBS.
10. Prepare samples for FACS sorting by staining the $CD3^-$ fraction with a viability dye and anti-CD56 antibody. Sample preparation for flow cytometry and cell sorting has been previously described in detail (Brummelman et al., 2019) (see *Note 15*).
11. APCs are sorted as $CD56^-$ and according to SSC-A and FSC-A parameters as depicted in the gating strategy in Fig. 2. Sorted APCs are

collected in R10 medium supplemented with IL-2 (40 IU/mL) and counted. Adjust cell volume to obtain a concentration of 10^5 cells/100 μL.

12. Fraction 2 is split in four fractions to separately enrich naïve and memory $CD4^+$ and $CD8^+$ T-cell subsets. Both naïve and memory fractions are subdivided unequally: 2/3 are used for $CD8^+$ enrichment and 1/3 for $CD4^+$ enrichment to compensate for the lower abundance of $CD8^+$ T cells in the peripheral blood; T cells are magnetically enriched with the EasySep™ T-cell enrichment kits (#19555, #19258, #19157, #19159 STEMCELL Technologies™) by following the manufacturer's instructions. Obtained cells are washed in PBS−/− by centrifuging at 500 × *g* for 10 min and counted (see *Note 16, 17*). $CD4^+$ and $CD8^+$ T cells are pooled together in a 1:1 ratio in 15 mL Falcon tubes for both naïve and memory subsets (see *Notes 18, 19*).
13. Enriched naïve and memory T-cell subsets ($CD4^+$ and $CD8^+$ pooled) are pelleted by centrifugation at 500 × *g* for 10 min and are stained with CFSE (see *Note 20*).
14. CFSE stock solution is diluted to a final concentration of 2 μM in PBS−/−, pre-warmed at 37 °C and added continuously to cells while vortexing to ensure homogenous staining. 1 mL is used to stain 5×10^6 cells; a minimum of 200 μL is used to stain as few as 5×10^3 cells.
15. Cells are incubated for 10 min at 37 °C in the dark.
16. After incubation, reaction is stopped by adding cold R10. The CFSE solution should be diluted at least 10-fold with R10.
17. Cells are pelleted by centrifugation at 500 × *g* for 10 min and supernatant is discarded. Resuspend cells in 10 mL R10.
18. Repeat step 17.
19. Cells are resuspended in 1–2 mL R10 and counted. Adjust volume to obtain a concentration of 10^5 cells/100 μL.
20. For each donor, add 100 μL of both the APC and naïve or memory T-cell suspensions to U-bottom 96 well plates, as follows (see *Note 21*):
    - Autologous APCs + naïve
    - Autologous APCs + memory
    - Third party (alloreactive) APCs + naïve
    - Third party (alloreactive) APCs + memory
    - Positive control: naïve or memory T cells + IL2 (20 IU/mL) + anti-CD3/CD28 microbeads resuspended in 100 μL R10 (see *Note 22*)
21. Cells are incubated for 5 days at 37 °C, 5% CO_2.

22. At the end of the incubation period, cells are harvested in 5 mL FACS tubes, washed with 4 mL PBS−/− by centrifugation at 500 × *g* for 10 min and resuspended in PBS−/−.
23. Cells are prepared to stain with custom panels for flow cytometric analysis (see step 10 for procedure).
24. Optional. Alloreactive T cells are FACS-sorted on the basis of CFSE dilution and CD25 expression according to the gate in Fig. 3A. CFSE undiluted, $CD25^-$ cells are sorted as control.

4. Notes

1. We provide catalog numbers of reagents and instrumentations used in our lab, but equivalent products with similar performance are available from multiple companies.
2. Blood samples from blood banks, even if tested for harmful viruses such as HIV and HCV, should be considered hazardous and must thus be manipulated according to BSL-2 safety procedures.
3. Antibodies used in this protocol are generally available as conjugated to several different fluorochromes. The user may choose fluorochromes that can be excited and detected by the flow cytometer and the cell sorter that are available in the lab/institutional facility.
4. Anti-APC or anti-FITC microbeads are also available from the same vendor. Accordingly, choose APC/FITC or APC tandem-conjugated antibodies for $CD3^+$ T-cell depletion.
5. Different flow cytometry platforms support different softwares for data acquisition.
6. Several providers offer alternative software options for the analysis of flow cytometry data.
7. Experiments described in this protocol used cells previously frozen in liquid nitrogen. Using fresh cells would result in higher viability, although we do not expect differences in the fraction of responding cells.
8. We generally obtain 2×10^6 PBMC/mL of whole blood and 150–500×10^6 PBMCs from buffy coats, depending on the donor. On the basis of our experience, the user should start from at least 150×10^6 PBMCs to obtain the minimum number of cells to successfully complete the protocol.
9. If cell clumps are evident, filter cell suspension with a 40 μm strainer.

10. We have noticed that the length of resting decreases background proliferation following incubation of T cells with autologous APCs (Fig. 3), nevertheless long resting (>24 h) might affect the survival of specific memory subsets, e.g., effector memory T cells that are more prone to apoptosis in the long term in the absence of cytokine support (Lugli et al., 2013a).
11. All antibodies should be carefully titrated with dedicated 2-fold serial dilution experiments. In general, unstimulated PBMCs can be used to test the majority of antibodies. However, in some instances, antigens might not be sufficiently expressed ex vivo, requiring overnight stimulation with anti-CD3/CD28. Detailed procedures related to antibody titration have been previously reported (Lugli et al., 2017).
12. The final volume of the reaction depends on the total number of cells to be sorted. In general, adding the predetermined concentration of the antibody to a final volume of 100 μL is generally optimal to stain up to 10^7 cells. If more cells are available, scale up accordingly. In general, we stain up to 10^8 cells in a final volume of 300 μL with 2–3 times the concentration of the antibody. Preliminary experiments with the target cell number should be performed to ensure that all cells are stained in the specific conditions.
13. Anti-CD3 antibodies can induce some degree of background activation. In our case, we observed an increased background with clone HIT-3A when compared to UCHT1.
14. Should PE-conjugated antibodies not be available, anti-PE microbeads can also bind cells stained with PE-Cy5 and PE-Cy7-conjugated antibodies. In general, we observed more efficient depletion when using PE-Cy7-conjugated antibodies. We suggest to perform preliminary tests to optimize the procedure.
15. Magnetic cell sorting can deplete >85% of $CD3^+$ T cells. FACS-based sorting is generally advised to isolate APCs to purity and remove $CD56^+$ NK cells. Alternatively, FACS-based sorting could be obviated by staining cells with anti-CD3 and anti-CD56 PE, followed by magnetic depletion with anti-PE microbeads, thus shortening the isolation procedure.
16. The current protocol was designed to test the alloreactive response of bulk naïve and memory T cells. Otherwise, specific cellular subsets could be purified with a cell sorter to test their alloreactive responses. Alternatively, FACS sorting could be used to enrich specific populations based on combinations of surface markers.

17. In the peripheral blood of healthy individuals, the CD4:CD8 ratio is generally 2:1. The PBMC preparation should be split according to the number of $CD4^+$ and $CD8^+$ T cells that are needed for the experiment.

18. The percentage of naïve T cells decreases with age, while that of memory T cells increases (Lugli et al., 2007). Different individuals might have completely different abundance of naïve and memory T cells in the peripheral blood, thus potentially biasing isolation towards one or the other subset. Percentages of naïve and memory T cells could be quantified by investigating CCR7 and CD95 expression by flow cytometry, as previously described (Lugli et al., 2013b). Bulk naïve are defined as $CCR7^+CD95^-$, bulk memory as $CD95^+$.

19. A minimum number of 2×10^5 naïve and memory T cells is needed for the successful completion of the experiment. Ensure to exceed this number to compensate for pipetting error.

20. Wash cells thoroughly in PBS−/− to ensure proteins are removed completely. The presence of leftover proteins from serum could interfere with CFSE staining.

21. We describe a protocol starting from two unrelated donors where T cells from each donor are incubated with APCs from the other donor. If starting from more donors, multiple allogeneic APC:T-cell combinations are possible.

22. Should naïve or memory cells not be sufficient after isolation, use PBMC from the same donors or from a third party donor as a positive control.

5. Concluding remarks

The protocol described here enables the fast and pure isolation of naïve and memory $CD4^+$ and $CD8^+$ T cells responding to third party APCs. The responding cells can be identified by CFSE dilution and activation markers expression, such as CD25. The responding cells can be isolated for downstream applications, such as functional assays (e.g., cytokine production following restimulation) or genomics (e.g., RNA-seq, ATAC-seq). We provide a protocol to obtain pure populations of APCs without contamination from $CD3^+$ T cells and NK cells, which could kill allogeneic naïve and memory T cells. The isolation procedure can be completed in about 1 day followed by the co-culture period of 5 days. We also provide an alternative strategy to obviate the need for FACS-based sorting of APCs, thus shortening the isolation procedure of about 3 h.

Acknowledgments

We wish to thank Eloise Scamardella, Simona Feno, Sabina Testa and Veronica Zanon for helping with cell isolation and antibody titrations. This work has been supported by a grant from the Ministero della Salute—Giovani Ricercatori GR-2016-02363315.

Conflicts of interest statement

E.L. has been holding research contracts with Bristol Myers Squibb on a topic unrelated to this manuscript and has received consulting/advisory honoraria from BD Biosciences, Biolegend, Swarm Therapeutics and Menarini. All other authors have no conflicts to declare.

References

Bleakley, M., et al. (2015). Outcomes of acute leukemia patients transplanted with naive T cell-depleted stem cell grafts. *The Journal of Clinical Investigation, 125*, 2677–2689.

Brummelman, J., et al. (2019). Development, application and computational analysis of high-dimensional fluorescent antibody panels for single-cell flow cytometry. *Nature Protocols, 14*, 1946–1969.

Castagna, L., et al. (2021). Feasibility and efficacy of CD45RA+ depleted donor lymphocytes infusion after haploidentical transplantation with post-transplantation cyclophosphamide in patients with hematological malignancies. *Transplantation and Cellular Therapy, 27*(478), e1–478.e5.

Eckle, S. B. G., Rossjohn, J., & McCluskey, J. (2013). Alloreactivity. In A. A. Zachary, & M. S. Leffell (Eds.), *Transplantation immunology: Methods and protocols* (pp. 3–39). Totowa, NJ: Humana Press. https://doi.org/10.1007/978-1-62703-493-7_1.

Felix, N. J., & Allen, P. M. (2007). Specificity of T-cell alloreactivity. *Nature Reviews. Immunology, 7*, 942–953.

Ferrara, J. L. M., & Chaudhry, M. S. (2018). GVHD: Biology matters. *Blood Advances, 2*, 3411–3417.

Foster, A. E., et al. (2004). Human CD62L- memory T cells are less responsive to alloantigen stimulation than CD62L+ naive T cells: Potential for adoptive immunotherapy and allodepletion. *Blood, 104*, 2403–2409.

Fowler, D. H. (2006). Shared biology of GVHD and GVT effects: Potential methods of separation. *Critical Reviews in Oncology/Hematology, 57*, 225–244.

Gratwohl, A., et al. (2017). Alloreactivity: The Janus-face of hematopoietic stem cell transplantation. *Leukemia, 31*, 1752–1759.

Lugli, E., Zanon, V., Mavilio, D., & Roberto, A. (2017). FACS analysis of memory T lymphocytes. *Methods in Molecular Biology (Clifton, N.J.), 1514*, 31–47.

Lugli, E., et al. (2007). Subject classification obtained by cluster analysis and principal component analysis applied to flow cytometric data. *Cytometry. Part J: the Journal of the International Society for Analytical Cytology, 71*, 334–344.

Lugli, E., et al. (2013a). Superior T memory stem cell persistence supports long-lived T cell memory. *The Journal of Clinical Investigation, 123*, 594–599.

Lugli, E., et al. (2013b). Identification, isolation and in vitro expansion of human and nonhuman primate T stem cell memory cells. *Nature Protocols, 8*, 33–42.

Macedo, C., et al. (2009). Contribution of naïve and memory T-cell populations to the human alloimmune response. *American Journal of Transplantation: Official Journal of the American Society of Transplantation and the American Society of Transplant Surgeons, 9*, 2057–2066.

Mahnke, Y. D., Brodie, T. M., Sallusto, F., Roederer, M., & Lugli, E. (2013). The who's who of T-cell differentiation: Human memory T-cell subsets. *European Journal of Immunology, 43*, 2797–2809.

Roberto, A., et al. (2015). Role of naive-derived T memory stem cells in T-cell reconstitution following allogeneic transplantation. *Blood*, *125*, 2855–2864.

Shook, D. R., et al. (2015). Haploidentical stem cell transplantation augmented by CD45RA negative lymphocytes provides rapid engraftment and excellent tolerability. *Pediatric Blood & Cancer*, *62*, 666–673.

Siu, J. H. Y., Surendrakumar, V., Richards, J. A., & Pettigrew, G. J. (2018). T cell Allorecognition pathways in solid organ transplantation. *Frontiers in Immunology*, *9*, 2548.

Touzot, F., et al. (2015). CD45RA depletion in HLA-mismatched allogeneic hematopoietic stem cell transplantation for primary combined immunodeficiency: A preliminary study. *The Journal of Allergy and Clinical Immunology*, *135*(1303–1309), e1–e3.

Warren, E. H., et al. (2012). Effect of MHC and non-MHC donor/recipient genetic disparity on the outcome of allogeneic HCT. *Blood*, *120*, 2796–2806.

CHAPTER SIX

Assessment of pDCs functional capacity upon exposure to tumor-derived soluble factors

Vladimír Koucký[a,b], Linn A. Syding[a], Klára Plačková[a,b], Lucie Pavelková[a,b], and Anna Fialová[a,*]

[a]Sotio, Prague, Czech Republic
[b]Department of Otorhinolaryngology and Head and Neck Surgery, First Medical Faculty, Motol University Hospital, Prague, Czech Republic
[*]Corresponding author: e-mail address: fialova@sotio.com

Contents

Abstract

Plasmacytoid dendritic cells (pDCs) are a minority subset of dendritic cells that despite their tiny quantity play an important role in the immune system, especially in antiviral immunity. They are known mostly as the major producers of type I IFN, which they secrete upon stimulation of endosomal Toll-like receptors 7 and 9 with viral RNA and DNA. However, the functionality of pDCs is more complex, as they were shown to be also involved in autoimmunity, inflammation, and cancer. In the context of the tumor microenvironment, pDCs mostly show substantial functional defects and thus contribute to establishing immunosuppressive micromilieu. Indeed, tumor-infiltrating pDCs were shown to be predominantly pro-tumorigenic, with reduced ability to produce IFNα and capacity to prime regulatory T cells via the ICOS/ICOS-L pathway.

Methods in Cell Biology, Volume 189
ISSN 0091-679X
https://doi.org/10.1016/bs.mcb.2024.07.002

Here we describe in detail a method to assess the functional capacity of pDCs upon exposure to tumor-derived cell culture supernatants. The same technique can be implemented with minimal variations to test any soluble factor's impact on pDC phenotype and function.

1. Introduction

Plasmacytoid dendritic cells (pDCs) account for less than 1% of peripheral blood mononuclear cells (PBMCs) in healthy individuals. They can be generated from common myeloid and lymphoid progenitors (Shortman et al., 2013). After development in the bone marrow, immature pDCs are released into peripheral blood and can directly enter secondary lymphatic organs and peripheral tissues, including pathologic conditions such as inflammatory and autoinflammatory lesions and tumors (Koucký et al., 2019).

Plasmacytoid DCs are mainly studied in relation to viral infections and autoimmune diseases and have been rather overlooked in immuno-oncology thus far. Indeed, pDCs are known as the most potent producers of interferon α (IFNα) from all peripheral blood cells (Siegal et al., 1999). The primary stimuli for type I IFN secretion are viral nucleic acids or self-derived DNA (Gilliet et al., 2008). Nucleic acids interact with Toll-like receptors 7 and 9 (TLR7, TLR9), resting in endosomes. After triggering TLRs, the pDC-derived IFN type I production depends on a complex containing MyD88 and IRF7. Within the complex, IRAK1 and/or IKKα phosphorylates IRF7, which is then translocated into the nucleus and regulates the expression of IFNα and other type I IFNs, namely, IFNβ, IFNτ and IFNω (Kawasaki & Kawai, 2014). The continuous high expression of IRF7 is specific for pDCs and allows them to secrete substantial amounts of IFNα and β rapidly (Dai et al., 2004). However, type I IFNs are neither the only cytokines produced by pDCs nor their only mode of interaction with other cells in the tumor microenvironment. Indeed, the MyD88-IRAK4-TRAF6 complex drives in pDCs NF-κB-dependent production of inflammatory cytokines, such as IL-6, IL-10, TNFα and IFNγ (Ito et al., 2006; McKenna et al., 2005). Besides cytokine production, the expression of chemokines CCL3, CCL4, CCL5, CXCL9, and CXCL10 has been reported in pDCs (Guillerey et al., 2012; Sozzani et al., 2010). Additionally, pDCs were also shown to be capable of presenting antigens (Tel et al., 2012), inducing Tregs (Martin-Gayo et al., 2010), and directly interacting with NK cells and B cells (Ito et al., 2006; Martin-Gayo et al., 2010; Tel et al., 2012; Young et al., 2008).

Originally, human pDCs were characterized as $CD4^+$ $CD123^+$ cells negative for lineage markers of B cells, T cells, NK cells, and monocytes (Grouard et al., 1997). Since then, more specific surface markers have been described (Wculek et al., 2019). A C-type lectin blood dendritic cell antigen 2 (BDCA2) is a commonly used marker for pDC identification. Unfortunately, engagement of this receptor with an antibody leads to a decrease in pDC functional capacity (Dzionek et al., 2001). Therefore, a blood dendritic cell antigen 4 (BDCA4) seems to be a better option for pDC enrichment (Dzionek et al., 2000). Standard markers used to identify pDCs in mice are B220, Ly6C, BST2, Siglec-H, and CD11c (Swiecki & Colonna, 2015).

Tumor-infiltrating pDCs were shown to be mostly defective in their functions in both humans and mice and were assumed to support the immunosuppressive setting of the tumor microenvironment (Vermi et al., 2011). Indeed, pDCs were identified as a negative prognostic marker in breast, ovarian, and oral cancers (Han et al., 2017; Labidi-Galy et al., 2012; Treilleux et al., 2004). In the tumor microenvironment, pDCs were shown to induce Tregs through the ICOS/ICOS-L pathway and by affecting L-tryptophan metabolism via indoleamine 2,3-dioxygenase (IDO) expression (Aspord et al., 2013; Conrad et al., 2012; Wculek et al., 2019). Depletion of pDCs led to a decrease in tumor burden and prolonged survival in breast cancer and glioma mouse models (Dey et al., 2015; Sawant et al., 2012). On the contrary, CpG ODN or TLR7L-stimulated pDCs were reported to be capable of lysing tumor cells in vitro and of reducing tumor burden in an experimental mouse model (Le Mercier et al., 2013; Wu et al., 2017).

However, given that pDCs are a rather sparse population in the tumor tissue, only a limited number of studies have analyzed the function of pDCs isolated directly from human tumors. Here we detail an accurate protocol for the isolation and treatment of circulating pDCs (Koucký et al., 2021). The functional capacity of pDCs was assessed as IFNα production upon stimulation with CpG ODN 2216 (TLR9 agonist) or with imiquimod (TLR7 agonist). Levels of IFNα were analyzed by flow cytometry and a commercially available ELISA kit.

2. Materials

2.1 Common disposables

- Nunc™ Cell Culture/Petri Dishes (#150340, Thermo Fisher Scientific)
- Screw Cap Microcentrifuge Tubes, 1.5 mL (#3467TS, Thermo Fisher Scientific)

- VACUETTE® Tubes 9 mL K2E K2EDTA (#455045, Greiner Bio-One)
- Nunc™ 50 mL Conical Sterile Polypropylene Centrifuge Tubes (#339652, Thermo Fisher Scientific)
- Corning™ Costar™ 96-Well, Cell Culture-Treated, U-Shaped-Bottom Microplates, individually wrapped (#10360691, Thermo Fisher Scientific)
- Cell strainer (#83.3945.100, Sarstedt)
- FACS tubes (e.g., Falcon® Round Bottom High-clarity Polypropylene Tubes, #352008, Corning)
- MACS LD columns (#130-042-901, Miltenyi Biotec)
- MASC MS columns (#130-042-201, Miltenyi Biotec)

2.2 Reagents

- Collagenase D (#11088866001, Roche)
- DNAse I (#10104159001, Roche)
- eBioscience™ Brefeldin A Solution (1000×) (#00-4506-51, Thermo Fisher Scientific)
- eBioscience™ Intracellular Fixation & Permeabilization Buffer Set (#88-8824-00, Thermo Fisher Scientific)
- PBS + EDTA (Gibco)
- Ficoll Paque™ Plus (GE HealthCare)
- Diamond Plasmacytoid Dendritic Cell Isolation Kit II, human (#130-097-240, Miltenyi Biotec)
- MACS buffer (PBS, 0.5% bovine serum albumin (BSA), and 2 mM EDTA)
- Complete RPMI 1640 medium: RPMI Medium 1640 (−)L-glutamine (Gibco) supplemented with 10% FBS, penicillin 1 U/mL + streptomycin 1 μg/mL (Penicillin-Streptomycin solution, Gibco) and Glutamax 20 μM (Gibco)
- CpG ODN 2216 (#tlrl-2216, InvivoGen)
- Imiquimod R837 (#tlrl-imqs-1, InvivoGen)
- Human IFNα ELISA kit (#41100-1, PBL Assay Science)

2.3 Equipment

- Benchtop cell counter, such as Countess II automated Cell Counter (Invitrogen)

- Benchtop centrifuge, such as Heraeus Megafuge 16R (Thermo Fisher Scientific)
- Humidified cell culture incubator, such as Heracell™ VIOS 160i CO_2 Incubator (Thermo Fisher Scientific)
- Laboratory biosafety cabinet (Class II), such as HERAsafe KS (Thermo Fisher Scientific)
- Flow cytometer, such as BD Fortessa
- MiniMACS Separator (Miltenyi Biotec)
- MidiMACS Separator (Miltenyi Biotec)
- Microplate reader, such as Spark® Multimode Microplate Reader (Tecan)

2.4 Software

- FlowJo v. 10.6 (FlowJo LLC)

3. Methods

3.1 Preparation of tumor-derived cell culture supernatants

1. Tumor tissue is transferred with sterile tweezers to a petri dish and cut into small pieces with sterile scissors
2. 5 mL of RPMI 1640 medium is added together with 1 mg/mL of collagenase D and 0.05 mg/mL DNase I
3. The petri dish is incubated on a shaker for 30 min at 37 °C under a gentle rocking motion
4. A sterile nylon 100 μm cell strainer is put on a 50 mL conical tube and rinsed with 5 mL of PBS
5. The minced tumor tissue is collected with a 10 mL pipette and transferred to the strainer
6. The tissue is passed through the strainer using a plunger of an insulin syringe and the strainer is rinsed with 2×5 mL of PBS
7. Collected cells are centrifuged at 1200 rpm for 10 min at 4 °C
8. The supernatant is discarded and cells are resuspended in 500 μL of PBS and counted using a benchtop cell counter
9. 20 mL of complete RPMI 1640 medium is added and cells are centrifuged at 1200 rpm for 10 min at 4 °C
10. The supernatant is discarded and cells are diluted to a concentration of 1×10^6 cells/mL

11. Cells are seeded in a 96-well U-bottom plate (200 μL of cell suspension/well ~2×10^5 of cells)
12. Cells are incubated for 24 h at 37 °C in 5% CO_2
13. After 24 h of incubation, the plate is centrifuged at 2200 rpm for 2 min, cell culture supernatants are collected in screw cap microcentrifuge tubes and frozen in −80 °C

3.2 pDC Isolation

1. Peripheral blood is collected in 9 mL K2E Vacuette tubes
2. To isolate mononuclear cells (PBMCs), fresh peripheral blood is diluted 1:1 with PBS + EDTA (see Note 1)
3. 30 mL of diluted blood is carefully layered onto 15 mL of Ficoll Paque™ Plus in a 50 mL conical tube
4. Diluted peripheral blood is centrifuged at 1800 rpm for 30 min at 20 °C under acceleration 1 and deceleration 0 settings (see Note 2)
5. The PBMC layer, located at interphase, is carefully collected with a pipette and transferred into a new conical 50 mL tube
6. Isolated PBMCs are resuspended in 50 mL of PBS + EDTA and centrifuged at 1200 rpm for 10 min at 4 °C
7. To remove platelets, the supernatant is discarded, and cells are resuspended in 50 mL PBS + EDTA and centrifuged at 1000 and 800 rpm, respectively, for 10 min at 4 °C
8. The supernatant is discarded and cells are resuspended in 500 μL of PBS
9. Isolated PBMCs are counted and subsequently centrifuged at 1200 rpm for 10 min
10. The supernatant is completely aspirated
11. To enrich pDCs, human Diamond Plasmacytoid Dendritic Cell Isolation Kit II (see Note 3) is used
12. The cell pellet is resuspended in 400 μL of MACS buffer per 10^8 total cells
13. 100 μL of Non-PDC Biotin-Antibody Cocktail II per 10^8 total cells is added and cells are incubated for 10 min at 4 °C
14. Cells are washed with 5 mL of MACS buffer, centrifuged at 1200 rpm for 10 min and the supernatant is completely aspired
15. Cell pellet is resuspended in 400 μL of MACS buffer per 10^8 total cells
16. 100 μL of Non-PDC MicroBead Cocktail II per 10^8 total cells is added and cells are incubated for 15 min at 4 °C

17. Cells are washed with 5 mL of MACS buffer, centrifuged at 1200 rpm for 10 min and the supernatant is completely aspired
18. The cell pellet is resuspended in 500 μL of MACS buffer per 10^8 total cells
19. Cells are separated using MACS LD column
20. Unlabeled cells (pre-enriched pDC fraction) that pass through the column are collected
21. Cells are centrifuged at 1200 rpm for 10 min and the supernatant is completely aspired
22. Cells are directly resuspended in 100 μL CD304 (BDCA-4/ Neuropilin-1) Diamond MicroBeads and incubated for 15 min at 4 °C
23. Cells are washed with 2 mL of MACS buffer, centrifuged at 1200 rpm for 10 min and the supernatant is completely aspired
24. The cell pellet is resuspended in 500 μL of MACS buffer per 10^8 total cells
25. Cells are separated using MACS MS column
26. Unlabeled cells that pass through the column are collected and discarded
27. The MS column is removed from the separator and placed on a 15 mL centrifuge tube
28. 1 mL of MACS buffer is added to the column and the magnetically labeled cells are immediately flushed out by firmly pushing the plunger into the column
29. Cells are washed with 5 mL of complete RPMI, centrifuged at 1200 rpm for 10 min and the supernatant is aspired
30. The cell pellet is resuspended in 500 μL of complete RPMI and cells are counted (see Note 4)

3.3 pDC seeding and treatment for ELISA

1. Enriched pDCs obtained at step 30 of Section 3.2 are diluted to a final concentration of 1×10^6 cells/mL in complete RPMI 1640
2. 50 μL of cell suspension ($\sim 5 \times 10^4$ cells) is pipetted in a 96-well U-bottom plate
3. 50 μL of tumor-derived cell culture supernatant is added to treated wells
4. 50 μL of complete RPMI 1640 medium is added to two control wells
5. Except for one control well, cells are stimulated with 5 μg/mL CpG ODN 2216 (prediluted in complete RPMI 1640 medium)
6. Cells are incubated for 24 h at 37 °C in 5% CO_2 (see Note 5)

7. After 24 h of incubation, the plate is centrifuged at 2200 rpm for 2 min and cell culture supernatants are collected
8. Levels of IFNα are measured using an IFNα ELISA kit (see Note 6)
9. 100 μL of the interferon standard, blank or sample are added to each well, the plate is covered with a sealer and incubated for 1 h
10. Supernatants are discarded and wells are washed with Wash Buffer
11. 100 μL of diluted Antibody Solution is added to each well, the plate is covered with a sealer and incubated for 1 h
12. Supernatants are discarded and wells are washed 3× with Wash Buffer
13. 100 μL of diluted HRP Solution is added to each well, the plate is covered with sealer and incubated for 1 h
14. Supernatants are discarded and wells are washed 4× with Wash Buffer
15. 100 μL of the TMB Substrate Solution is added to each well, plate is incubated in the dark at room temperature for 15 min
16. 100 μL of Stop Solution is added to each well
17. Absorbance is determined at 450 nm within 5 min after the addition of the Stop Solution

3.4 pDC seeding and treatment for flow cytometry

1. Enriched pDCs obtained at step 30 of Section 3.2 are diluted to a final concentration of 1×10^6 cells/ml in complete RPMI 1640 medium
2. 50 μL of cell suspension (~5×10^4 cells) is pipetted in a 96-well U-bottom plate
3. 50 μL of tumor-derived cell culture supernatant is added to treated wells
4. 50 μL of complete RPMI 1640 medium is added to two control wells
5. Except for one control well, cells are stimulated with 5 μg/mL CpG ODN 2216 (prediluted in complete RPMI 1640 medium) or 5 μg/mL imiquimod (prediluted in complete RPMI 1640 medium) (see Note 7)
6. Cells are incubated for 15 h at 37 °C in 5% CO_2 (see Notes 5 and 8)
7. After 2 h of incubation, 3.0 μg/mL of Brefeldin A is added to each well
8. After 15 h of incubation, cells are transferred to a 96V-shaped well plate and centrifuged at 2200 rpm for 2 min
9. The supernatant is discarded and cells are washed with 200 μL of PBS
10. Cells are stained with panel of primary antibodies and fixable viability dye: anti-CD3 FITC, anti-CD14 FITC, anti-CD16 FITC, anti-CD19 FITC, anti-CD20 FITC, anti-CD56 FITC (all "Lineage" antibodies),

Table 1 Antibodies used for flow cytometry staining.

Antigen	Fluorochrome	Clone	Producer
Fixable viability dye	eFluor 780		eBioscience
CD3	FITC	OKT3	eBioscience
CD4	PE/Dazzle 594	OKT4	Biolegend
CD14	FITC	MEM-15	Exbio
CD16	FITC	LNK16	Exbio
CD19	FITC	LT19	Exbio
CD20	FITC	LT20	Exbio
CD45	Alexa Fluor 700	HI30	Biolegend
CD56	FITC	MEM-188	Exbio
CD123	APC	6H6	Biolegend
BDCA-2	PerCP-Cy 5.5	201A	Biolegend
IFNα	PE	7N4–1	BD Biosciences

anti-CD4 PE/Dazzle 594, anti-CD45 AF700, anti-CD123 APC, anti-BDCA-2 PerCP-Cy5.5 and eBioscience™ Fixable Viability Dye eFluor™ 780 (Table 1) in PBS + 2% FBS for 20 min at 4 °C

11. Cells are washed with PBS
12. 100 μL of fixation buffer is added and prepared as follows: Fixation/Permeabilization Concentrate and Fixation/Permeabilization Diluent in a 1:3 ratio
13. The plate is incubated for 20 min at 4 °C
14. 100 μL of permeabilization buffer, diluted in distilled water in a 1:9 ratio is added and the plate is centrifuged at 2200 rpm for 2 min
15. The supernatant is discarded and 150 μL of permeabilization buffer is added and the plate is centrifuged at 2200 rpm for 2 min
16. The supernatant is discarded and cells are stained with anti-IFNα PE in permeabilization buffer for 20 min at 4 °C
17. 150 μL of permeabilization buffer is added and the plate is centrifuged at 2200 rpm for 2 min
18. Supernatants are discarded, 200 μL of PBS is added and the plate is centrifuged at 2200 rpm for 2 min

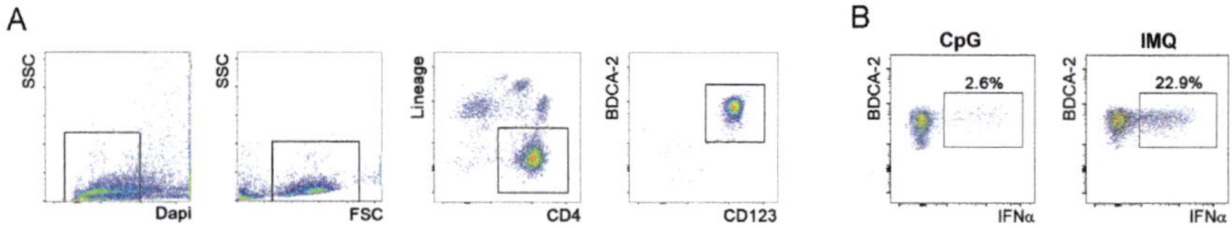

Fig. 1 (A) Gating strategy for assessing the purity of pDCs enriched from PBMCs. Dapi$^-$ live cells were selected, debris was excluded, Lineage$^-$CD4$^+$ cells were selected and, as a last step, BDCA-2$^+$CD123$^+$ pDCs were gated. The purity was assessed as the proportion of BDCA-2$^+$CD123$^+$ pDCs among live cells. (B) Detection of IFNα expression in pDCs upon stimulation with 5 μg/mL CpG ODN 2216 and 5 μg/mL imiquimod (IMQ) for 15 h. Enriched pDCs cultivated in complete RPMI 1640 in the presence of Brefeldin A are shown.

19. Supernatants are discarded and cells are resuspended in 100 μL of PBS + 2% FBS
20. Cells are analyzed by flow cytometry (examples of results observed in control cells are shown in Fig. 1)

4. Notes

1. Peripheral blood should be as fresh as possible, after 5 h, the functional capacity of pDCs rapidly declines.
2. Use a centrifuge with adjustable acceleration and deceleration settings. The rotor brakes must be inactivated entirely to effectively prevent mixing of the phases.
3. Keep cells and buffers cold, work fast. The detailed protocol provided by the manufacturer can be found at: https://static.miltenyibiotec.com/asset/150655405641/document_tkkslfg3ld6sdf6vvamc984s3q?content-disposition=inline
4. The purity of pDCs obtained by this method was 70–80% when gated on live cells. The initial proportion of pDCs in PBMCs was 0.1–0.25%. The gating strategy of enriched pDCs is shown in Fig. 1A.
5. Cell viability could be enhanced by adding 10 ng/mL of IL3.
6. Detailed protocol provided by the manufacturer can be found at: https://www.pblassaysci.com/sites/default/files/document/2022-06/41100_protocol_rev.05_1.pdf
7. Imiquimod may reduce the viability of the cells.
8. Production of IFNα has different kinetics upon stimulation of CpG ODN and imiquimod. Imiquimod induces rapid production of IFNα

clearly detectable by intracellular flow cytometry after 5–6 h. On the contrary, the peak of IFNα expression in CpG ODN-treated pDCs is observed after 24 h of stimulation.

5. Concluding remarks

Plasmacytoid DCs, although infrequent, may play an important role in the immunosuppression of the tumor microenvironment. Here we describe a detailed protocol on how to employ peripheral blood-derived pDC to evaluate the effects of soluble factors (such as cytokines in tumor-derived single-cell suspension culture supernatants) on pDC functional capacity. As the production of type I IFNs is the hallmark of pDC function, we provide a protocol for detecting IFNα in cell culture supernatants and intracellularly in pDCs. Thus, the overall quantity of IFNα produced and the proportion of IFNα-producing pDCs can be assessed.

References

Aspord, C., et al. (2013). Plasmacytoid dendritic cells support melanoma progression by promoting Th2 and regulatory immunity through OX40L and ICOSL. *Cancer Immunology Research*, *1*(6), 402–415. https://doi.org/10.1158/2326-6066.CIR-13-0114-T.

Conrad, C., et al. (2012). Plasmacytoid dendritic cells promote immunosuppression in ovarian cancer via ICOS costimulation of Foxp3+ T-regulatory cells. *Cancer Research*, *72*(20), 5240–5249. https://doi.org/10.1158/0008-5472.CAN-12-2271.

Dai, J., et al. (2004). Regulation of IFN regulatory factor-7 and IFN-alpha production by enveloped virus and lipopolysaccharide in human plasmacytoid dendritic cells. *Journal of Immunology*, *173*(3), 1535–1548. https://doi.org/10.4049/jimmunol.173.3.1535.

Dey, M., et al. (2015). Dendritic cell-based vaccines that utilize myeloid rather than plasmacytoid cells offer a superior survival advantage in malignant glioma. *Journal of Immunology (Baltimore, Md.: 1950)*, *195*(1), 367–376. https://doi.org/10.4049/jimmunol.1401607.

Dzionek, A., et al. (2000). BDCA-2, BDCA-3, and BDCA-4: Three markers for distinct subsets of dendritic cells in human peripheral blood. *Journal of Immunology*, *165*(11), 6037–6046. https://doi.org/10.4049/jimmunol.165.11.6037.

Dzionek, A., et al. (2001). BDCA-2, a novel plasmacytoid dendritic cell-specific type II C-type lectin, mediates antigen capture and is a potent inhibitor of interferon alpha/beta induction. *The Journal of Experimental Medicine*, *194*(12), 1823–1834. https://doi.org/10.1084/jem.194.12.1823.

Gilliet, M., et al. (2008). Plasmacytoid dendritic cells: Sensing nucleic acids in viral infection and autoimmune diseases. *Nature Reviews. Immunology*, *8*(8), 594–606. https://doi.org/10.1038/nri2358.

Grouard, G., et al. (1997). The enigmatic plasmacytoid T cells develop into dendritic cells with interleukin (IL)-3 and CD40-ligand. *The Journal of Experimental Medicine*, *185*(6), 1101–1111. https://doi.org/10.1084/jem.185.6.1101.

Guillerey, C., et al. (2012). Pivotal role of plasmacytoid dendritic cells in inflammation and NK-cell responses after TLR9 triggering in mice. *Blood*, *120*(1), 90–99. https://doi.org/10.1182/blood-2012-02-410936.

Han, N., et al. (2017). Increased tumor-infiltrating plasmacytoid dendritic cells predicts poor prognosis in oral squamous cell carcinoma. *Archives of Oral Biology*, *78*, 129–134. https://doi.org/10.1016/j.archoralbio.2017.02.012.

Ito, T., et al. (2006). Specialization, kinetics, and repertoire of type 1 interferon responses by human plasmacytoid predendritic cells. *Blood*, *107*(6), 2423–2431. https://doi.org/10.1182/blood-2005-07-2709.

Kawasaki, T., & Kawai, T. (2014). Toll-like receptor signaling pathways. *Frontiers in Immunology*, *5*, 461. https://doi.org/10.3389/fimmu.2014.00461.

Koucký, V., et al. (2019). Immunology of plasmacytoid dendritic cells in solid tumors: A brief review. *Cancers*, *11*(4), 470. https://doi.org/10.3390/cancers11040470.

Koucký, V., et al. (2021). The cytokine milieu compromises functional capacity of tumor-infiltrating plasmacytoid dendritic cells in HPV-negative but not in HPV-positive HNSCC. *Cancer Immunology, Immunotherapy*, *70*(9), 2545–2557. https://doi.org/10.1007/s00262-021-02874-y.

Labidi-Galy, S. I., et al. (2012). Plasmacytoid dendritic cells infiltrating ovarian cancer are associated with poor prognosis. *Oncoimmunology*, *1*(3), 380–382. https://doi.org/10.4161/onci.18801.

Le Mercier, I., et al. (2013). Tumor promotion by intratumoral plasmacytoid dendritic cells is reversed by TLR7 ligand treatment. *Cancer Research*, *73*(15), 4629–4640. https://doi.org/10.1158/0008-5472.CAN-12-3058.

Martin-Gayo, E., et al. (2010). Plasmacytoid dendritic cells resident in human thymus drive natural Treg cell development. *Blood*, *115*(26), 5366–5375. https://doi.org/10.1182/blood-2009-10-248260.

McKenna, K., et al. (2005). Plasmacytoid dendritic cells: Linking innate and adaptive immunity. *Journal of Virology*, *79*(1), 17–27. https://doi.org/10.1128/JVI.79.1.17-27.2005.

Sawant, A., et al. (2012). Depletion of plasmacytoid dendritic cells inhibits tumor growth and prevents bone metastasis of breast cancer cells. *The Journal of Immunology*, *189*(9), 4258–4265. https://doi.org/10.4049/jimmunol.1101855.

Shortman, K., et al. (2013). Chapter four—Plasmacytoid dendritic cell development. *Advances in Immunology*, *120*, 105–126. https://doi.org/10.1016/B978-0-12-417028-5.00004-1.

Siegal, F. P., et al. (1999). The nature of the principal type 1 interferon-producing cells in human blood. *Science*, *284*(5421), 1835–1837. https://doi.org/10.1126/science.284.5421.1835.

Sozzani, S., et al. (2010). Trafficking properties of plasmacytoid dendritic cells in health and disease. *Trends in Immunology*, *31*(7), 270–277. https://doi.org/10.1016/j.it.2010.05.004.

Swiecki, M., & Colonna, M. (2015). The multifaceted biology of plasmacytoid dendritic cells. *Nature Reviews. Immunology*, *15*(8), 471–485. https://doi.org/10.1038/nri3865.

Tel, J., et al. (2012). Human plasmacytoid dendritic cells are equipped with antigen-presenting and tumoricidal capacities. *Blood*, *120*(19), 3936–3944. https://doi.org/10.1182/blood-2012-06-435941.

Treilleux, I., et al. (2004). Dendritic cell infiltration and prognosis of early stage breast cancer. *Clinical Cancer Research: An Official Journal of the American Association for Cancer Research*, *10*(22), 7466–7474. https://doi.org/10.1158/1078-0432.CCR-04-0684.

Vermi, W., et al. (2011). Plasmacytoid dendritic cells and cancer. *Journal of Leukocyte Biology*, *90*(4), 681–690. https://doi.org/10.1189/jlb.0411190.

Wculek, S. K., et al. (2019). Effective cancer immunotherapy by natural mouse conventional type-1 dendritic cells bearing dead tumor antigen. *Journal for Immunotherapy of Cancer*, *7*(1), 100. https://doi.org/10.1186/s40425-019-0565-5.

Wu, Q., et al. (2017). Breast cancer subtypes predict the preferential site of distant metastases: A SEER based study. *Oncotarget*, *8*(17), 27990–27996. https://doi.org/10.18632/oncotarget.15856.

Young, L. J., et al. (2008). Differential MHC class II synthesis and ubiquitination confers distinct antigen-presenting properties on conventional and plasmacytoid dendritic cells. *Nature Immunology*, *9*(11), 1244–1252. https://doi.org/10.1038/ni.1665.

CHAPTER SEVEN

Degranulation assay to evaluate NK cell natural and antibody-dependent cell-mediated cytotoxicity against A549 tumor spheroids

Ainara Lopez-Pardo[a], Ainhoa Amarilla-Irusta[a], Víctor Sandá[a], Mario Stan-Fontoba[a], Francisco Borrego[a,b], and Laura Amo[a,b,*]

[a]Immunopathology Group, Biobizkaia Health Research Institute, Barakaldo, Spain
[b]Ikerbasque, Basque Foundation for Science, Bilbao, Spain
[*]Corresponding author: e-mail address: laura.amoherrero@bio-bizkaia.eus

Contents

Abstract

Adoptive natural killer (NK) cell-based immunotherapy is a promising treatment approach in cancer that is showing notable efficacy against hematological malignancies. However, the success of NK cell immunotherapy in patients with solid tumors is limited due to several barriers, which include the immunosuppressive tumor

Methods in Cell Biology, Volume 189
ISSN 0091-679X
https://doi.org/10.1016/bs.mcb.2024.06.003

microenvironment (TME), heterogeneity of tumor cells and poor NK cell infiltration into the tumor. Advances in 3D *in vitro* culture technologies have opened new avenues for the development of more physiological human cancer models that mimic important tumor features which are absent in traditional 2D studies and may be essential for the improvement of immunotherapies against solid tumors. Here, we describe a comprehensive protocol to generate tumor spheroids from the A549 lung carcinoma cell line, then establish co-cultures with NK cells to, ultimately, determine NK cell functional response with a degranulation assay, a surrogate of NK cell cytotoxicity against tumor spheroids. Additionally, we studied degranulation by stimulating NK cell antibody-dependent cell-mediated cytotoxicity (ADCC) with cetuximab, an IgG1 monoclonal antibody used in cancer therapy. Likewise, other monoclonal antibodies or combination treatments could also be studied in this 3D co-culture system, providing very valuable information to define effective combinations of therapeutic agents able to generate NK cells with high cytotoxic potential that could lead to more successful adoptive NK cell-based therapies for the treatment of solid tumors.

1. Introduction

Natural killer (NK) cells are lymphocytes of the innate immune system classified as part of the family of the innate lymphoid cells. In humans, NK cells comprise between 5% and 20% of peripheral blood mononuclear cells (PBMCs) and are generally identified as CD3/14/19^{neg} lymphocytes expressing CD56, although different subtypes of human NK cells can be defined on the basis of the relative expression of CD16a (or FcγRIIIA) and CD56. CD7 is a cell surface receptor, which is expressed on lymphocytes, and it may also help identifying NK cells in combination with the absence of B cell and T cell markers, i.e., CD19 and CD3, respectively (Milush et al., 2009).

One of the fundamental roles that characterize NK cells is their ability to recognize and eliminate abnormal cells, such as viral infected or tumor cells, without prior sensitization. NK cells express activating receptors on their surface, including NKG2D, natural cytotoxicity receptors (NKp30, NKp44, and NKp46), NKp80, 2B4, CD94/NKG2C, activating KIR and DNAM-1, as well as inhibitory receptors, such as inhibitory KIR and CD94/NKG2A (Borrego et al., 2002; Borrego, Masilamani, Marusina, Tang, & Coligan, 2006; Burgess et al., 2008; Orrantia, Terrén, Astarloa-Pando, Zenarruzabeitia, & Borrego, 2021). The integrated signals of these receptors determine NK cell responses. Hence, in order to carry out their cytotoxic activity, effector and target cells need to interact through the activating NK cell receptors that recognize stress-induced ligands on damaged cells, and this interaction needs to surpass the signal mediated by the inhibitory receptors, resulting in NK cell activation and the subsequent killing of target cells. Also, NK cells can interact with opsonized cells, which are cells coated with IgG antibodies that bind to certain antigens on the surface of the target cell. Through the CD16a receptor, NK cells recognize the fragment crystallizable (Fc) portion of the IgG, leading to the activation of NK cells and the elimination of the target cells in a process known as antibody-dependent cell-mediated cytotoxicity (ADCC) (Lo Nigro et al., 2019). In fact, the ADCC mechanism is exploited in several therapeutic interventions, one of those being the use of monoclonal antibodies (mAbs) for targeted cancer therapy (Ochoa et al., 2017; Zahavi, AlDeghaither, O'Connell, & Weiner, 2018). Examples of mAbs in use are cetuximab, which targets the human epidemic growth factor receptor (EGFR) and it is widely used to treat colorectal cancer and other tumors (Bonner et al., 2006; Jonker et al., 2007); trastuzumab which targets the human epidemic growth factor receptor 2 (HER2) and it is employed for breast cancer (Slamon et al., 2001); and rituximab, which targets CD20 antigen (CD20) and it is used for some types of lymphoma and leukemia (Coiffier et al., 2002; Maury et al., 2016).

Once NK cells are activated, their lytic function is exerted with the release of cytolytic granules to the immunological synapse, causing the death of target cells. These granules contain perforin and granzymes. Perforin produces pores in the plasma membrane of the target cell, facilitating the entry of granzymes into it, which cause target cell apoptosis (Lopez et al., 2013). CD107a, or LAMP-1, is a membrane protein found in the cytolytic granules that is upregulated on the surface of NK cells following stimulation, and it is used as a functional marker of NK cell degranulation (Alter, Malenfant, & Altfeld, 2004).

Given the key role of NK cells in cancer immune surveillance, many NK cell-based therapies are currently being developed and employed in clinical trials. Some of these approaches include increasing the efficacy of the adoptive transfer of NK cells with cytokines, antibodies targeting activating receptors on NK cells and/or with immune checkpoint inhibitors that block NK cell inhibition. Although the adoptive transfer of NK cells has proved to be an effective treatment of hematopoietic cancers, preclinical and clinical data indicate that the efficacy of NK cell infusion in patients with solid malignancies is hindered by several factors, including the immunosuppressive tumor microenvironment (TME) and poor tumor infiltration of NK cells (Habif, Crinier, André, Vivier, & Narni-Mancinelli, 2019; Terrén, Orrantia, Vitallé, Zenarruzabeitia, & Borrego, 2019; Tong et al., 2022).

In order to better understand NK cell functionality in solid tumors and design improved NK cell therapies, three-dimensional (3D) cultures that more accurately replicate the *in vivo* tumor architecture have been developed. Depending on the source, establishment protocol of the 3D tumor cultures, the culturing method and their biological and architectural complexity, different 3D models have been defined, which include tumor organoids, spheroids, tumorspheres, and tumor-on-a-chip, among others. Tumor spheroids are one of the most common and easy-to-use scaffold-free models for 3D cell culture. They result from the aggregation of cancer cells *in vitro*, which self-organize into tumor microstructures recapitulating important tumor features, such as hypoxia, cell–cell cohesion and interaction, tumor morphology or nutrient gradients (Khaitan & Dwarakanath, 2006; Nunes, Barros, Costa, Moreira, & Correia, 2019; Zhao, Fong, Seow, & Toh, 2023). Given these properties, tumor spheroids provide a very valuable system for the *in vitro* evaluation of cancer immunotherapies, for assessing the efficiency of therapeutic antibodies and for the enhancement of immune cell infiltration and response against tumor spheroidal targets.

This protocol describes step-by-step the establishment of tumor spheroids from A549 lung carcinoma cell line, and their use to assess NK cell anti-tumor response. In this context, we present the methodology to test NK cell degranulation and its enhancement through ADCC stimulation induced by cetuximab.

2. Materials and equipment

2.1 Common disposables

- 0.22 μm filters (Corning, ref. 431097).
- 75-cm^2 flasks (Corning, ref. 431464U).
- 15 mL falcon and 50 mL conical tubes (Falcon, ref. 352096 & 352070).
- 96-well Clear Round Bottom Ultra-Low Attachment Microplate, Individually Wrapped, with Lid, Sterile (Corning, ref. 7007).
- 70 μm cell strainers (Falcon, ref. 352350).
- LS columns for magnetic cell sorting (Miltenyi Biotec, ref. 130-042-401).
- Flow cytometry tubes with cell strainer cap (Falcon, ref. 352235).
- 1.5 mL reaction tubes (Sarstedt, ref. 72.690.001).
- Filtered pipet tips multiple volumes (Teknovas, ref. 713264 & 713118; Neptune, ref. 713078N & BT300).
- Serological pipettes, 5 mL and 10 mL (Deltalab, ref. 900034.C & 900036.C).

2.2 Equipment

- Gilson pipettes.
- Pipetboy (Integra Biosciences, ref. 155521).
- In-VitroCell NU-5810 Direct Heat CO_2 Incubator (Nuaire).
- Neubauer chamber (Kisker, ref. 393000).
- Centrifuge 5810R (Eppendorf).
- Primovert Inverted Microscope (Zeiss).
- MACS™ MultiStand (Miltenyi Biotec, ref. 130-042-303).
- MidiMACS™ Separator (Miltenyi Biotec, ref. 130-042-302).
- LSRFortessa X-20 (BD Biosciences).
- Fiji Software.

2.3 Reagents

- Dulbecco's Modified Eagle Medium (DMEM, Corning, ref. 15-013-CV).
- Fetal Bovine Serum (FBS, Gibco, ref. A5256701).
- Penicillin-Streptomycin (P/S, 10,000 U/mL, Gibco, ref. 15140122).
- MEM Non-Essential Amino Acids Solution 100× (NEAA, Gibco, ref. 11140-035).
- Sodium Pyruvate 100× (Gibco, ref. 11360-070).
- Trypsin-EDTA solution (Gibco, ref. 25200056).
- Trypan Blue (Sigma-Aldrich, ref. T8154).
- Phosphate Buffered Saline (PBS) 1× (Corning, ref. 21-040-CV).
- Ficoll® Paque Plus (GE Healthcare, ref. 17-1440-02).
- Rosewell Park Memorial Institute (RPMI) 1640 with L-glutamine (Corning, ref. 10-040-CV).
- Ethylenediaminetetraacetic Acid (EDTA, Fisher Scientific, ref. BP2482-500).
- NK cell Isolation Kit (Miltenyi, ref. 130-092-657).
- Cetuximab—Erbitux 5 mg/mL (Merck).
- GolgiStop (BD Biosciences, ref. 554724).
- GolgiPlug (BD Biosciences, ref. 555029).
- Bovine Serum Albumin (BSA, Sigma-Aldrich, ref. A3733).
- Brilliant Stain Buffer (BD Biosciences, ref. 563794).

3. Methodology

3.1 A549 cell culture

3.1.1. A549 lung carcinoma cell line is grown in DMEM supplemented with 5% heat inactivated FBS (Note 5.1), 1% NEAA and 1% P/S

(Note 5.2). This culture media will be hereinafter referred to as cDMEM. A549 cells are cultured in the incubator at standard culture conditions (37 °C and 5% CO_2) in conventional 75-cm^2 flasks and are used within 10 passages.

3.1.2. Before full confluence is reached, cells are gently washed with 10 mL of PBS at room temperature (RT) and subsequently detached with 1.5 mL of Trypsin-EDTA solution. The reaction is stopped by adding 8 mL of cDMEM to the cells. Cells are collected, centrifuged at 300×*g* for 5 min and counted in a Neubauer chamber by the Trypan Blue exclusion method (Note 5.3).

3.2 Quality pre-assessment of A549 spheroids

Prior to the generation of A549 spheroids for the coculture with NK cells, the optimal A549 cell seeding number is determined by plating different number of cells, ranging from 500 to 10^4 cells per well in 100 μL of cDMEM on a 96-Well Ultra-Low Attachment Microplate.

From the quality pre-assessment of the spheroids, we conclude that the ideal cell seeding number for the co-cultures is 10^4 cells per well, as it complies with our spheroid quality criteria as well as with the specific cell number requirements for performing the NK cell degranulation assay. The parameters analyzed to reach this conclusion are the following:

3.2.1. **Spheroid growth and integrity**: Phase contrast images are acquired at 10× magnification from day 0 (A549 cell seeding day) to day 7 in the 96-well plate. The micrographies evidence that A549 cell line successfully generates spheroids. In fact, within 48 h of culture, spheroids generated from 10^4 cells evolve from an initial loose clustering of cells to the spheroidal state, i.e., compact round morphology with a defined exterior boundary resulting from cell-to-cell tight adhesions (Fig. 1A and B).

3.2.2. **Spheroid size:** Micrographies from day 0 to 7 are further analyzed using Fiji software by measuring the area (A) of each spheroid. Spheroid diameters (d) are then calculated as follows: $d=2\times\sqrt{((A)/(\pi))}$. The results indicate that the diameter for A549 spheroids generated from 10^4 cells keeps around 500–600 μm from day 1 to 7. These values are suitable for our assay since tumor spheroids of around 500 μm have proven useful for the study of solid tumors as they generate nutritional and oxygen gradients that mimic their native counterparts (Fig. 1C and D, Note 5.4, Mehta, Hsiao, Ingram, Luker, & Takayama, 2012).

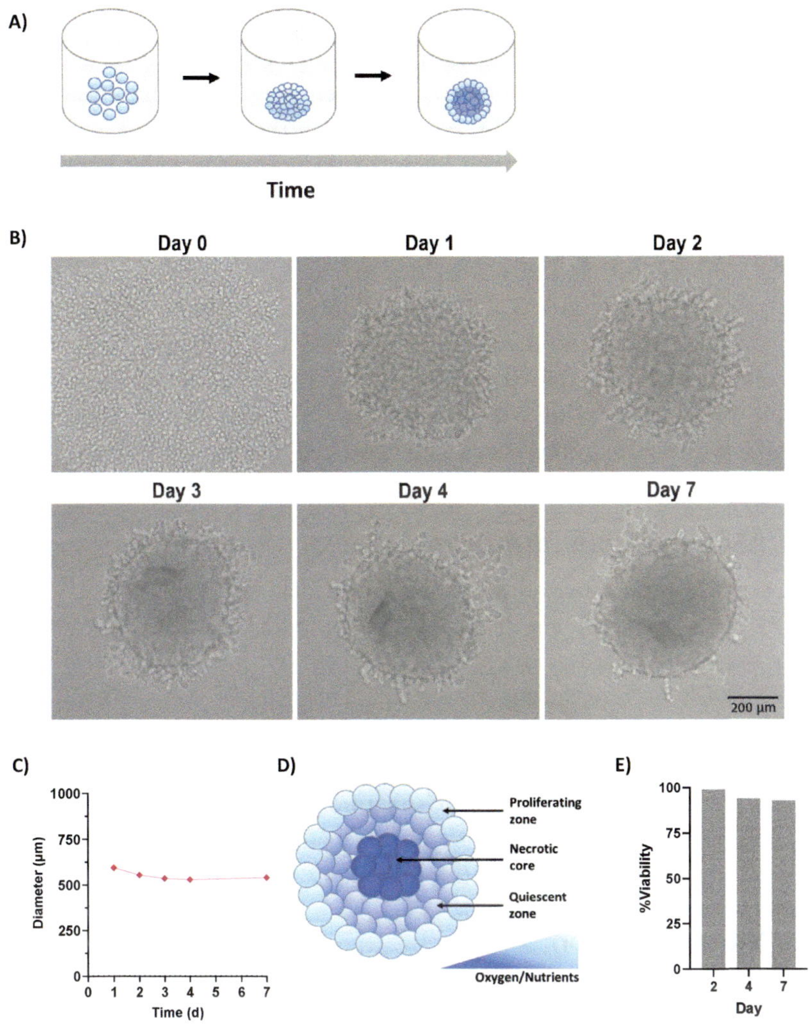

Fig. 1 Quality pre-assessment of A549 spheroids. (A) Schematic representation of spheroid formation initiated from loose cell aggregates that compact and reach a solid spheroidal state over time. (B) Representative 10× magnification micrographs of A549 spheroid formation initiated from 10^4 cells, which show the compaction of the spheroids over the time and the formation of the exterior boundary of the spheroid. Scale bar 200 μm. (C) Evolution of the diameter of a representative A549 spheroid generated from 10^4 cells. (D) Schematic representation of a tumor spheroid, which consists of sequential layers of proliferating cells, quiescent cells and a necrotic core due to the oxygen and nutrient gradients. (E) Viability at days 2, 4 and 7 of a representative A549 spheroid generated from 10^4 cells.

3.2.3. **Viability:** As a result of the oxygen and nutrient gradients, tumor spheroids commonly consist of a proliferating ring of cells that surrounds a quiescent interior, and often present a pathophysiological necrotic core (Fig. 1D, Note 5.5). On the other hand, cells at the spheroid boundary may begin to dissociate from the main body. Thus, viability needs to be addressed by the Trypan Blue exclusion method after disaggregating the spheroids with Trypsin-EDTA (the collection and dissociation of the spheroids will be described in detail in Section 3.6.1). The viability of A549 spheroids originated from 10^4 cells is over 92% from day 0 to 7 (Fig. 1E).

In conclusion, the initial seeding of 10^4 A549 cells gives rise to compact round spheroids showing adequate integrity, size and growth kinetics that comply with the requirements for the NK cell degranulation assay.

3.3 A549 spheroid formation

3.3.1. To induce spheroid formation for the degranulation assay, 10^4 A549 cells per well are plated in 100 µL of cDMEM on a 96-Well Ultra-Low Attachment Microplate (Note 5.6). For this assay, six wells per condition are seeded. The plate is then centrifuged at $300 \times g$ for 1 min so that cells settle at the bottom of the plate.

3.3.2. Cells are cultured at 37 °C and 5% CO_2 for 48 h and checked daily to determine the formation of spheroids, i.e. compacted cell aggregate with a defined exterior boundary, before proceeding to the subsequent functional assays (Note 5.7).

3.4 NK cell isolation

3.4.1. PBMCs are obtained from buffy coats by density gradient centrifugation using Ficoll® Plaque Plus following the methodology described in Terrén, Orrantia, Vitallé, Zenarruzabeitia, and Borrego (2020) (Note 5.8). Commonly, over 400×10^6 PBMCs are obtained from a single buffy coat.

3.4.2. NK cells are then isolated from the PBMCs suspension by magnetic cell sorting using the NK Cell Isolation Kit from Miltenyi, following manufacturer's instructions. Briefly, up to 200×10^6 PBMCs are incubated with NK cell Biotin-Antibody Cocktail for 5 min at 4 °C, followed by 10 min incubation with NK Cell MicroBead Cocktail, and are subjected to a magnetic negative selection of

NK cells using LS columns. The number of NK cells obtained is approximately 20×10^6.

3.4.3. NK cell purity is determined by flow cytometry based on CD56 expression and lack of CD3, CD14, CD19 and CD123 markers. For that, specific fluorochrome-conjugated antibodies are used as shown in Table 1. NK cell purity is, generally, over 90%.

3.4.4. Once NK cells are isolated, they are rested overnight (16–18 h) in the incubator in RPMI 1640 with L-glutamine and supplemented with 10% FBS, 1% NEAA, 1% Sodium Pyruvate and 1% P/S. This culture media will be hereinafter referred to as cRPMI. Specifically, NK cells are plated at a concentration of 10^6 cells/mL in 3 mL of cRPMI per well in a 6-well plate.

3.5 Co-culture of spheroids and NK cells: Degranulation assay

Once NK cells are isolated and rested, and A549 spheroids have reached the solid spheroidal state, the co-culture and functional assays are performed under sterile conditions.

3.5.1. Spheroids and NK cells are co-cultured at 1:1 E:T ratio (based on the initial seeding number of A549 cells). 10^4 NK cells are added in cRPMI media to the tumor spheroids in the 96-well plate reaching a final volume of 200 μL per well (Note 5.9).

3.5.2. To induce ADCC, 1 μg/mL of cetuximab per well is added to the corresponding conditions.

3.5.3. For assessing NK cell degranulation, 0.4 μL of FITC-labeled anti-CD107a is added per well. From this step onwards, protection from light is required.

3.5.4. The co-cultures are incubated for 1 h at 37 °C and 5% CO_2.

3.5.5. To stop vesicle trafficking, 0.66 μL/mL of GolgiStop (monensin) and 1 μL/mL of GolgiPlug (brefeldin A) are added per well (Note 5.10).

3.5.6. Co-cultures are incubated overnight (16–18 h) at 37 °C and 5% CO_2.

3.6 Flow cytometry

For assessing NK cell degranulation after the overnight co-cultures, spheroids need to be first dissociated to obtain a single cell suspension required for flow cytometry.

Table 1 Antibodies and fluorescent dyes employed for flow cytometry analyses.

Laser	Fluorochrome	Marker	Manufacturer	Ref.	Clone	Volume added per sample (μL)	Total volume per sample (μL)
405 nm	BV421	CD56	BD Biosciences	562751	NCAM16.2	3	100
	BV510	CD3	BD Biosciences	563109	UCHT1	3	100
	BV510	CD14	BD Biosciences	563079	MφP9	3	100
	BV510	CD19	BD Biosciences	562947	SJ25C1	3	100
	BV510	CD123	BD Biosciences	563072	9F5	3	100
488 nm	FITC	CD107a	Miltenyi Biotec	130-111-620	REA792	0.4	200
	PE	CD7	BD Biosciences	555361	M-T701	20	100
635 nm	Near-IR	LIVE/DEAD	Invitrogen	L34976	NA	1	1000

3.6.1 Sample collection

3.6.1.1. To collect the samples for their analysis, cells from each condition are transferred from 6 wells into 15 mL Falcon tubes and they are topped up with PBS (Note 5.11). Samples are centrifuged at 300×*g* for 5 min to pellet the cells at the bottom of the tube.

3.6.1.2. The supernatant is removed with extreme care, as spheroids are easily disturbed from the bottom of the tube and could be aspirated together with the supernatant.

3.6.1.3. Spheroids are dissociated by incubating with 200 μL of Trypsin-EDTA solution at 37 °C for approximately 15 min (Note 5.12). When microscope observation of the 15 mL Falcon tube evidences that spheroids have been disrupted into single cells (Note 5.13), the trypsinization process needs to be stopped by adding 10 μL of FBS.

3.6.1.4. Samples are topped up with PBS and centrifuged at 300×*g* for 5 min at RT. The pellet is resuspended in 1 mL of PBS after discarding the supernatant (Note 5.14).

3.6.1.5. Samples are filtered into flow cytometry tubes using their cell strainer caps by adding 1 mL of PBS and centrifuging at 300×*g* for 5 min afterwards.

3.6.2 Stainings

3.6.2.1. For viability staining, cell pellets are resuspended in 1 mL of PBS, and 1 μL of reconstituted LIVE/DEAD Fixable Near-IR reagent is added per sample. Cells are then incubated for 30 min at 4 °C protected from light (Note 5.15).

3.6.2.2. Cells are washed two times by adding 2 mL of 2.5% BSA-PBS and centrifuging at 300×*g* for 5 min at 4 °C. The supernatant is discarded afterwards.

3.6.2.3. For extracellular staining, 50 μL of Brilliant Stain Buffer and 50 μL of 2.5% BSA-PBS is added to each tube together with the specific fluorochrome-conjugated anti-human antibodies (anti-CD3, CD7, CD14, CD19, and CD123) in the volumes described in Table 1. Cells are incubated for 30 min at 4 °C in the dark (Notes 5.15–5.17).

3.6.2.4. Cells are washed two times by adding 2 mL of 2.5% BSA-PBS and centrifuging at 300×*g* for 5 min at 4 °C. The supernatant is discarded afterwards.

3.6.2.5. Lastly, cells are resuspended in 200 μL PBS prior to sample acquisition in the BD LSRFortessa X-20 cytometer (Note 5.18).

3.7 Analysis of NK cell degranulation

In this section, we will analyze the outcome of a degranulation assay performed with human isolated NK cells in the presence of tumor spheroids, and we will additionally assess whether the addition of cetuximab to the co-cultures results in increased degranulation of NK cells.

The data is analyzed using FlowJo v.10.7.1 software, although other analysis softwares could be employed as well. In order to analyze degranulation, we need to select first the population of NK cells. For this purpose, we follow the next gating strategy (Fig. 2):

1. First, a time gating is applied, e.g., plotting the side scatter area (SSC-A) vs time, to exclude possible artifacts that can happen during the acquisition (Fig. 2A).
2. Lymphocytes are gated based on the forward and side scatter area (FSC-A and SSC-A, respectively) parameters (Fig. 2B).
3. Doublet cells or clumps can significantly affect the analysis leading to inaccurate conclusions, so two doublet exclusions are performed consecutively: SSC-A vs SSC-H (height) density plot first (Fig. 2C), followed by FSC-A vs FSC-H (Fig. 2D).
4. Then, dead cells are excluded by selecting LIVE/DEAD negative cells (Fig. 2E).
5. Finally, NK cells are identified based on the negative expression of CD3, CD14, CD19, CD123 and positive expression of CD7 (Fig. 2F, Note 5.16).

Once NK cells are gated, we analyze the percentage of degranulating NK cells by determining the percentage of CD107a positive cells in different conditions: (1) NK cells cultured alone reveals the NK cell basal degranulation, (2) spheroid co-culture condition discloses the spheroid-induced degranulation, and (3) the addition of cetuximab to the co-cultures shows the increase of degranulation induced by ADCC. In our assay, the percentage of cells expressing CD107a gives us clear information on the behavior of NK cells in the different conditions, i.e., exposure to A549 spheroids induces NK cell degranulation, which sharply increases in the presence of cetuximab as it triggers ADCC in NK cells (Fig. 3A). These results obtained by flow cytometry analysis are in line with the behavior observed under the microscope (Fig. 3B), which shows that after 16–18 h of NK cell-A549 spheroid co-culture, the integrity of the spheroid is notably affected, as evidenced by the loss of the exterior boundary and the presence of a cloud of debris and dying cells surrounding the spheroid, which is exacerbated with the addition of cetuximab to the co-cultures. In contrast, the integrity

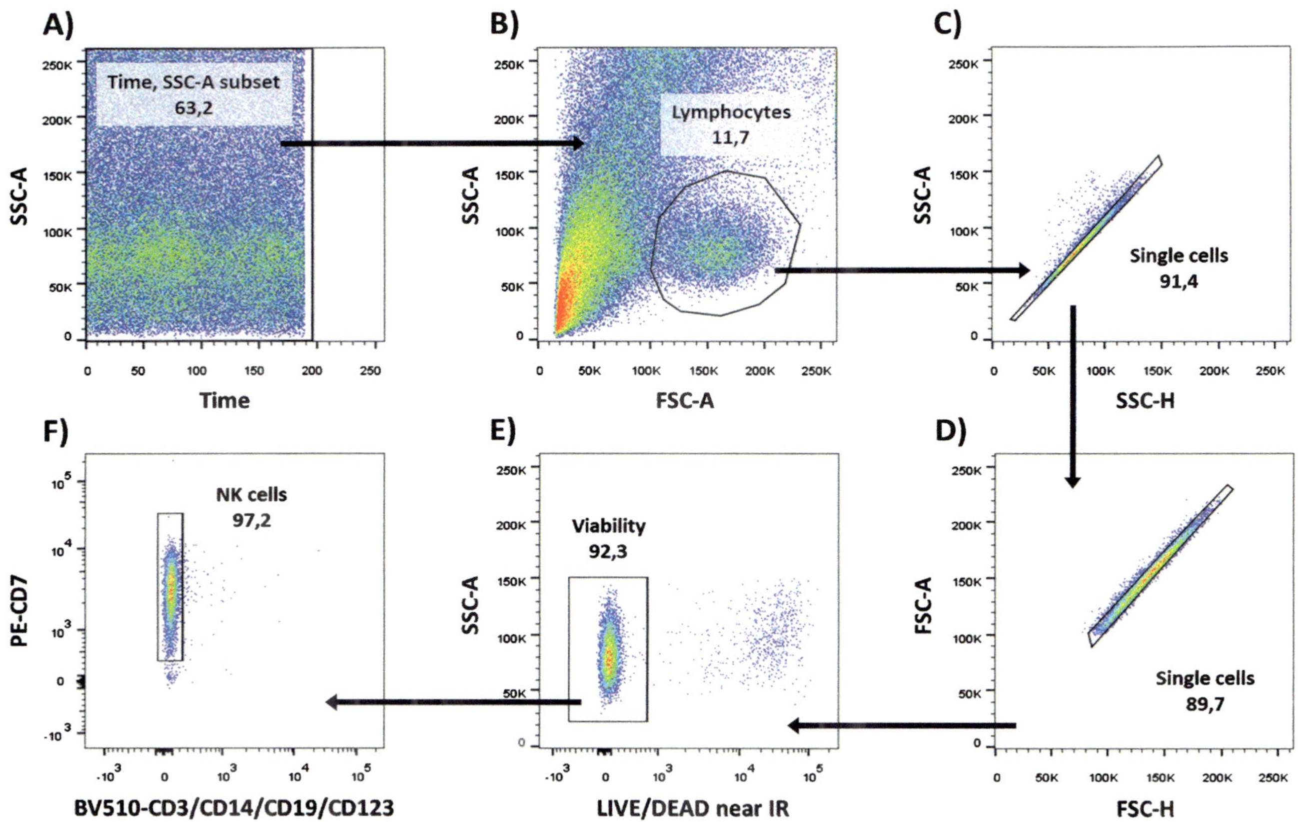

Fig. 2 Gating strategy of NK cells cultured with A549 spheroids. Pseudocolor plots obtained by FlowJo v.10.7.1 software. (A) First, the data is refined using a "Time Gate," plotting SSC-A vs time, which allows selecting areas with a stable flow stream. (B) Then lymphocytes are identified based on the FSC-A and SSC-A parameters. (C and D) Single cells are gated by two consecutive doublet exclusions: SSC-A vs SSC-H (C), followed by FSC-A vs FSC-H (D). (E) Viable cells are selected as they are negative for LIVE/DEAD dye staining. (F) Lastly, NK cells are identified as they are positive for CD7, but they do not express CD3, CD14, CD19, and CD123 markers.

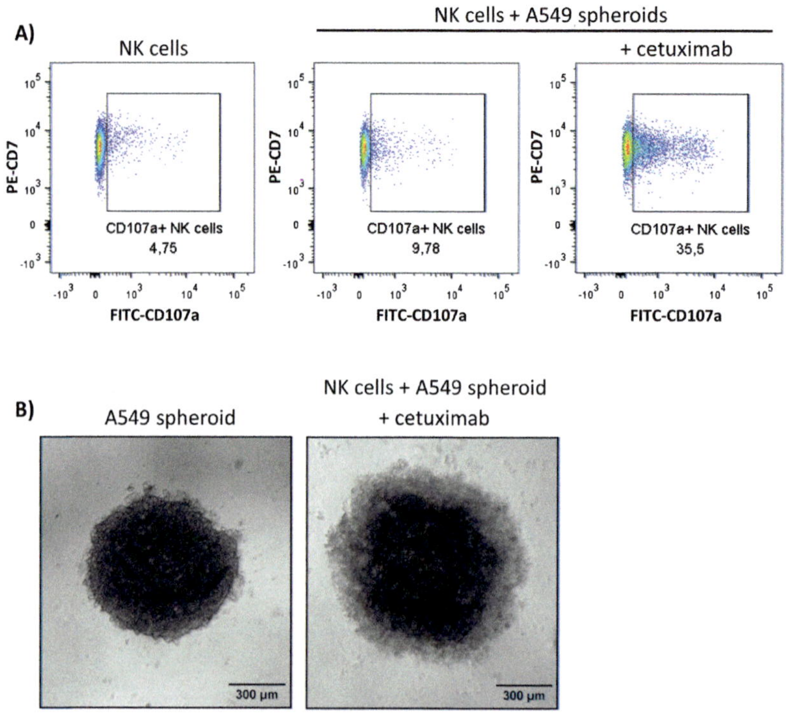

Fig. 3 NK cell degranulation assessment after the co-culture with A549 spheroids. (A) Pseudocolor plots showing the percentage of NK cells that degranulate (CD107a+) in three different conditions: NK cells cultured with and without A549 spheroids (1:1 E:T ratio) and 1 µg/mL of cetuximab for 16–18 h. (B) Phase contrast representative images at 4× magnification of an A549 spheroid cultured alone or with NK cells plus cetuximab, which evidence the tumor spheroid integrity loss in the co-culture condition. Scale bar 300 µm.

of spheroids cultured alone is unaltered after the 16–18 h incubation time. This suggests that NK cells in the presence of tumor spheroids are active and capable of exerting cytotoxic activity against tumor spheroids, which is enhanced by the addition of cetuximab.

4. Concluding remarks

Analyzing the interaction of NK cells with tumor spheroids is of great interest since clinical success with immunotherapy for solid malignancies is proving to be quite challenging. Therefore, spheroids, which closely mimic some of the biological features of solid tumors, provide us with valuable data on NK cell infiltration and killing that could be used to ultimately improve

NK cell-based immunotherapies. In this context, the assay described above provides a reliable and quantifiable method to assess whether NK cells respond to tumor spheroids as they do when exposed to tumor cell lines in 2D cultures. Specifically, it allows us to evaluate if tumor spheroids induce NK cell degranulation. Since NK cells use the ADCC mechanism to recognize and kill cells expressing target antigens when they are antibody-coated, monoclonal antibodies such as cetuximab or trastuzumab can also be included in the assay to check if the activation of NK cells is augmented in their presence. This assay can also be performed with tumor spheroids from different cell line origins after doing an initial assessment of spheroid growth to adapt cell numbers and spheroid incubation time as required. In conclusion, although additional work is required, our 3D studies using tumor spheroids suggest that immunotherapy against solid malignancies could be potentially benefited from combining NK cell infusion with therapeutic antibodies such as cetuximab. This combination could improve NK cell anti-tumor responses in patients with solid tumors.

5. Notes

5.1. FBS is heat-inactivated by incubation at 56 °C for 30 min in a water bath.

5.2. The culture media is filtered with 0.22 μm strainers and stored at 4 °C up to a month.

5.3. Trypan Blue stains cells with compromised cell membranes (i.e., dead cells). Count only viable (not Trypan Blue-stained) cells.

5.4. Tumor spheroids starting from 200 μm of diameter present chemical gradients (oxygen, nutrients and catabolites), and develop a central necrotic area from a diameter of 500 μm onwards. Cells located in the spheroid periphery reflect the *in vivo* situation of actively cycling tumor cells adjacent to capillaries while innermost cells become quiescent and eventually die. Depending on the object of the study, the size of the spheroid would need to be adjusted to better mimic the biological tumor context. In order to do so, there are two main factors to take into account: first, the higher the number of cells initially seeded, the larger the size of the resulting spheroid, and, second, the culture time of the spheroid also allows to grow larger spheroids until they reach their maximum size (Han, Kwon, & Kim, 2021; Singh et al., 2020; Zanoni et al., 2016).

5.5. The growth of the necrotic core in spheroids positively correlates with the size of the given spheroid, which depends on the initial seeding number and the culture time.

5.6. To generate spheroids from different tumor cell lines, the optimal seeding concentration would need to be assessed in each case since it varies depending on the growth rate and cell type.

5.7. In our hands, A549 spheroids reach spheroidal morphology approximately 48 h after the initial seeding of 10^4 cells. It is important to note that the passage number and the confluence of A549 cells prior to cell harvesting for spheroid formation, together with variations in cDMEM preparation, among others, may influence the time of spheroid formation.

5.8. For PBMCs collection, the buffy coat is diluted (1:4) in sterile PBS. 10 mL of the dilution is added over 5 mL of Ficoll® Paque Plus in 15 mL conical tubes. Tubes are centrifuged at 800×*g* for 20 min at RT with low acceleration and brake. The PBMCs layer is then collected, transferred into a single 50 mL conical tube and washed with sterile PBS. PBMCs are washed two times by centrifuging at 300×*g* for 5 min at RT and filtered with 70 μm cell strainers to avoid cell clumps on the cell suspension.

5.9. For this degranulation assay, we employed the following three conditions: NK cells cultured alone, NK cells cultured with spheroids, and NK cells cultured with spheroids and cetuximab. For each condition, six replicates were cultured and combined in order to obtain an adequate number of NK cells for flow cytometry analysis. Also, additional NK cells were cultured for unstained controls for flow cytometry analysis.

5.10. GolgiStop and GolgiPlug should be stored at 4 °C. We recommend bringing them to RT during the step 3.5.4.

5.11. We recommend using a 1000 μL pipette for spheroid collection. Visually inspect that the spheroids have been transferred to the tubes and do not remain attached to the bottom of the plate wells (the size of the spheroids at this stage allows direct eye visualization).

5.12. Before the trypsinization, it would be ideal to wash the samples with PBS, but due to the high risk of losing the spheroids in the processing, we directly add trypsin-EDTA to each sample (after very carefully removing as much supernatant as possible).

5.13. During the trypsinization process, every 5 min mix the co-cultures with a 200 μL pipette to aid the dissociation of the spheroids. It is also of great importance to check the dissociation progress under the brightfield microscope to subject the cells to the shortest trypsinization time possible. Thus, as soon as spheroids are no longer visible under the microscope and, instead, a single cell suspension is observed, proceed to stop the trypsinization.

5.14. After step 3.6.1.4, it is possible to stop the protocol and store the sample tubes at 4 °C to proceed to the staining on the following day.

5.15. Vortex the tubes to ensure proper mixing of the cells and dyes or antibodies before incubating them.

5.16. In our NK cell-A549 spheroid co-cultures, we have employed anti-CD7 for NK cell identification instead of the canonical CD56 (which we use routinely with 2D assays).

5.17. We recommend titrating the antibody dilution for an optimal staining. Information about the antibody volumes used in this experiment can be found in Table 1.

5.18. We strongly recommend performing a compensation with single-stained controls prior to data acquisition.

Acknowledgments

We thank Javier Díez Garcia and Miren Solaun from the microscopy and flow cytometry platforms of Biobizkaia Health Research Institute. This study was supported by the following grant: Department of Health, Basque Government (2022333018). Ainara Lopez-Pardo is recipient of a predoctoral contract funded by "la Caixa" Foundation (ID 100010434; LCF/BQ/DI22/11940012). Ainhoa Amarilla-Irusta is recipient of a predoctoral contract funded by Fundación AECC-Spanish Association Against Cancer (PRDVZ234209AMAR). Ainara Lopez-Pardo, Ainhoa Amarilla-Irusta and Víctor Sandá are recipients of a fellowship from the Jesús de Gangoiti Barrera Foundation (FJGB23/007, FJBG21/005 and FJGB23/005, respectively). Laura Amo is an Ikerbasque Research Fellow and Francisco Borrego is an Ikerbasque Research Professor, Ikerbasque, Basque Foundation for Science.

Conflicts of interests

The authors declare that the research was conducted in the absence of any commercial or financial relationships that could be construed as a potential conflict of interest.

References

Alter, G., Malenfant, J. M., & Altfeld, M. (2004). CD107a as a functional marker for the identification of natural killer cell activity. *Journal of Immunological Methods*, *294*(1–2), 15–22. https://doi.org/10.1016/j.jim.2004.08.008.

Bonner, J. A., Harari, P. M., Giralt, J., Azarnia, N., Shin, D. M., Cohen, R. B., et al. (2006). Radiotherapy plus cetuximab for squamous-cell carcinoma of the head and neck. *New England Journal of Medicine*, *354*(6), 567–578. https://doi.org/10.1056/NEJMoa053422.

Borrego, F., Kabat, J., Kim, D. K., Lieto, L., Maasho, K., Peña, J., et al. (2002). Structure and function of major histocompatibility complex (MHC) class I specific receptors expressed on human natural killer (NK) cells. *Molecular Immunology*, *38*(9), 637–660. https://doi.org/10.1016/s0161-5890(01)00107-9.

Borrego, F., Masilamani, M., Marusina, A. I., Tang, X., & Coligan, J. E. (2006). The CD94/NKG2 family of receptors: From molecules and cells to clinical relevance. *Immunologic Research*, *35*(3), 263–278. https://doi.org/10.1385/IR:35:3:263.

Burgess, S. J., Maasho, K., Masilamani, M., Narayanan, S., Borrego, F., & Coligan, J. E. (2008). The NKG2D receptor: Immunobiology and clinical implications. *Immunologic Research*, *40*(1), 18–34. https://doi.org/10.1007/s12026-007-0060-9.

Coiffier, B., Lepage, E., Briere, J., Herbrecht, R., Tilly, H., Bouabdallah, R., et al. (2002). CHOP chemotherapy plus rituximab compared with CHOP alone in elderly patients with diffuse large-B-cell lymphoma. *New England Journal of Medicine*, *346*(4), 235–242. https://doi.org/10.1056/NEJMoa011795.

Habif, G., Crinier, A., André, P., Vivier, E., & Narni-Mancinelli, E. (2019). Targeting natural killer cells in solid tumors. *Cellular & Molecular Immunology*, *16*, 415–422. https://doi.org/10.1038/s41423-019-0224-2.

Han, S. J., Kwon, S., & Kim, K. S. (2021). Challenges of applying multicellular tumor spheroids in preclinical phase. *Cancer Cell International*, *21*, 152. https://doi.org/10.1186/s12935-021-01853-8.

Jonker, D. J., O'Callaghan, C. J., Karapetis, C. S., Zalcberg, J. R., Tu, D., Au, H. J., et al. (2007). Cetuximab for the treatment of colorectal cancer. *New England Journal of Medicine*, *357*(20), 2040–2048. https://doi.org/10.1056/NEJMoa071834.

Khaitan, D., & Dwarakanath, B. S. (2006). Multicellular spheroids as an in vitro model in experimental oncology: Applications in translational medicine. *Expert Opinion on Drug Discovery*, *1*, 663–675. https://doi.org/10.1517/17460441.1.7.663.

Lo Nigro, C., Macagno, M., Sangiolo, D., Bertolaccini, L., Aglietta, M., & Merlano, M. C. (2019). NK-mediated antibody-dependent cell-mediated cytotoxicity in solid tumors: Biological evidence and clinical perspectives. *Annals of Translational Medicine*, 7(5), 105. https://doi.org/10.21037/atm.2019.01.42.

Lopez, J. A., Susanto, O., Jenkins, M. R., Lukoyanova, N., Sutton, V. R., Law, R. H. P., et al. (2013). Perforin forms transient pores on the target cell plasma membrane to facilitate rapid access of granzymes during killer cell attack. *Blood*, *121*(14), 2659–2668. https://doi.org/10.1182/blood-2012-07-446146.

Maury, S., Chevret, S., Thomas, X., Heim, D., Leguay, T., Huguet, F., et al. (2016). Rituximab in B-lineage adult acute lymphoblastic leukemia. *New England Journal of Medicine*, *375*(11), 1044–1053. https://doi.org/10.1056/NEJMoa1605085.

Mehta, G., Hsiao, A. Y., Ingram, M., Luker, G. D., & Takayama, S. (2012). Opportunities and challenges for use of tumor spheroids as models to test drug delivery and efficacy. *Journal of Controlled Release*, *164*(2), 192–204. https://doi.org/10.1016/j.jconrel.2012.04.045.

Milush, J. M., Long, B. R., Snyder-Cappione, J. E., Cappione, A. J., 3rd, York, V. A., Ndhlovu, L. C., et al. (2009). Functionally distinct subsets of human NK cells and monocyte/DC-like cells identified by coexpression of CD56, CD7, and CD4. *Blood*, *114*(23), 4823–4831. https://doi.org/10.1182/blood-2009-04-216374.

Nunes, A. S., Barros, A. S., Costa, E. C., Moreira, A. F., & Correia, I. J. (2019). 3D tumor spheroids as in vitro models to mimic in vivo human solid tumors resistance to therapeutic drugs. *Biotechnology and Bioengineering*, *116*, 206–226. https://doi.org/10.1002/bit.26845.

Ochoa, M. C., Minute, L., Rodriguez, I., Garasa, S., Perez-Ruiz, E., Inogés, S., et al. (2017). Antibody-dependent cell cytotoxicity: Immunotherapy strategies enhancing effector NK cells. *Immunology & Cell Biology*, *95*(4), 347–355. https://doi.org/10.1038/icb.2017.6.

Orrantia, A., Terrén, I., Astarloa-Pando, G., Zenarruzabeitia, O., & Borrego, F. (2021). Human NK cells in autologous hematopoietic stem cell transplantation for cancer treatment. *Cancers*, *13*(7), 1589. https://doi.org/10.3390/cancers13071589.

Singh, S. K., Abbas, S., Saxena, A. K., Tiwari, S., Sharma, L. K., & Tiwari, M. (2020). Critical role of three-dimensional tumorsphere size on experimental outcome. *BioTechniques*, *69*(5), 333–338. https://doi.org/10.2144/btn-2020-0081.

Slamon, D. J., Leyland-Jones, B., Shak, S., Fuchs, H., Paton, V., Bajamonde, A., et al. (2001). Use of chemotherapy plus a monoclonal antibody against HER2 for metastatic breast cancer that overexpresses HER2. *New England Journal of Medicine*, *344*, 783–792. https://doi.org/10.1056/NEJM200103153441101.

Terrén, I., Orrantia, A., Vitallé, J., Zenarruzabeitia, O., & Borrego, F. (2019). NK cell metabolism and tumor microenvironment. *Frontiers in Immunology*, *10*, 2278. https://doi.org/10.3389/fimmu.2019.02278.

Terrén, I., Orrantia, A., Vitallé, J., Zenarruzabeitia, O., & Borrego, F. (2020). CFSE dilution to study human T and NK cell proliferation in vitro. *Methods in Enzymology*, *631*, 239–255. https://doi.org/10.1016/bs.mie.2019.05.020.

Tong, L., Jiménez-Cortegana, C., Tay, A. H., Wickström, S., Galluzzi, L., & Lundqvist, A. (2022). NK cells and solid tumors: Therapeutic potential and persisting obstacles. *Molecular Cancer*, *21*, 206. https://doi.org/10.1186/s12943-022-01672-z.

Zahavi, D., AlDeghaither, D., O'Connell, A., & Weiner, L. M. (2018). Enhancing antibody-dependent cell-mediated cytotoxicity: A strategy for improving antibody-based immunotherapy. *Antibody Therapeutics*, *1*(1), 7–12. https://doi.org/10.1093/abt/tby002.

Zanoni, M., Piccinini, F., Arienti, C., Zamagni, A., Santi, S., Polico, R., et al. (2016). 3D tumor spheroid models for in vitro therapeutic screening: A systematic approach to enhance the biological relevance of data obtained. *Scientific Reports*, *6*, 19103. https://doi.org/10.1038/srep19103.

Zhao, J., Fong, A., Seow, S. V., & Toh, H. C. (2023). Organoids as an enabler of precision immuno-oncology. *Cells*, *12*(8), 1165. https://doi.org/10.3390/cells12081165.

CHAPTER EIGHT

Characterization of Natural Killer cells infiltrating irradiated murine tumors through flow cytometry

Jérémy Baude[a,b], Riad Ladjohounlou[b], Emeric Limagne[c,d,e,f], Lisa Froidurot[b], Véronique Morgand[b], and Céline Mirjolet[a,b,g,*]

[a]Department of Radiation Oncology, Centre Georges-François Leclerc, Dijon, France
[b]Preclinical Radiation Therapy and Radiobiology Unit, Centre Georges-François Leclerc, Dijon, France
[c]University of Burgundy, Dijon, France
[d]Cancer Biology Transfer Platform, Centre Georges-François Leclerc, Dijon, France
[e]Genetic and Immunology Medical Institute, Dijon, France
[f]INSERM LNC-UMR1231, Dijon, France
[g]TIReCS Team, UMR INSERM 1231, Dijon, France
*Corresponding author: e-mail address: cmirjolet@cgfl.fr

Contents

Abstract

The radiation-induced immune response is increasingly well documented. However, some aspects remain unclear, notably the role of Natural Killer (NK) cells, a subgroup of innate lymphoid cells involved in the antitumor response, in the response to RT.

ISSN 0091-679X
https://doi.org/10.1016/bs.mcb.2024.05.012

It therefore seems necessary to better characterize NK cells infiltrating irradiated tumors in order to better understand the mechanisms of action of RT, enabling its subsequent optimization and combination with other immunomodulatory treatments. A key technology for studying intratumoral immune cells is flow cytometry, which can simultaneously quantify and analyze the phenotype of numerous cells.

Here, we propose a method for phenotyping intratumoral NK cells through flow cytometry in mice bearing colorectal tumors treated with radiotherapy. This procedure can also be used to study the radiation-induced NK cell response in a wide range of solid tumors.

1. Introduction

Radiotherapy (RT) is central to the treatment of numerous cancers. Its efficacy has historically been attributed to its ability to induce multiple DNA damages leading to cell cycle arrest and eventually to cell death, especially in cancer cells that often show difficulties in repairing such damage (Bernier, Hall, & Giaccia, 2004).

In addition, an abundant literature has shown that RT is a powerful immunomodulatory treatment (Boustani, Grapin, Laurent, Apetoh, & Mirjolet, 2019). On one hand, it can reshape the tumor intro a pro-inflammatory environment through the release of danger-associated molecular patterns (DAMPs), tumor antigens (Lhuillier, Rudqvist, Elemento, Formenti, & Demaria, 2019) and cytokines such as interferon γ (IFN-γ) conducive to the attraction and activation of multiple effectors such as dendritic cells, $CD8^+$ T cells and M1 macrophages that participate in the antitumor activity of RT (Charpentier, Spada, Van Nest, & Demaria, 2022). On the other hand, RT can also provoke lymphopenia (Chen et al., 2020; Punjabi et al., 2021) and trigger the production of immunosuppressive cytokines such as transforming growth factor β (TGF-β) (De Martino, Daviaud, & Vanpouille-Box, 2021; Shevtsov, Sato, Multhoff, & Shibata, 2019; Vanpouille-Box et al., 2015) and the upregulation of immune checkpoints such as programmed cell death protein 1 (PD-1)-PD1-ligand (PD-L1) (Bardhan, Anagnostou, & Boussiotis, 2016; Dovedi & Illidge, 2015; Grapin et al., 2019; McLaughlin et al., 2020; Wang et al., 2022; Yi et al., 2022). This result in the inhibition of effector cells and the infiltration of immunosuppressive cells within the tumor microenvironment (TME) among which are regulatory T cells (Gao, Shi, & Wang, 2022; Liu et al., 2015; Tanchot et al., 2013) and myeloid-derived suppressive cells (MDSCs) (Kang, Jeong, Song, & Choi, 2020).

Recently, emerging evidence have supported the fact that Natural Killer (NK) cells are also modulated by ionizing radiations (Baude, Limagne, Ladjohounlou, & Mirjolet, 2023; Canter et al., 2017; Walle et al., 2018) and play a significant role in the antitumor efficacy of RT (Chiossone, Dumas, Vienne, & Vivier, 2018; Vivier et al., 2011). They represent a subset of innate lymphocytes that do not express polymorphic clonotypic receptors, but a large number of activating and inhibitory receptors whose balance of activation drives cell activity (Pegram, Andrews, Smyth, Darcy, & Kershaw, 2011). Activated NK cells induce an antigen-independent lyse of cancer cells through (1) the release of perforin and granzyme B in the vicinity of target cells (Vivier, Tomasello, Baratin, Walzer, & Ugolini, 2008), (2) an antibody-dependent cell-mediated cytotoxicity (ADCC) (Lo Nigro et al., 2019), (3) apoptosis *via* their expression of fas death receptor ligand (FASL) and tumor-necrosis-factor related apoptosis inducing ligand (TRAIL) death domains (Vivier et al., 2008) and (4) the production of pro-inflammatory tumor necrosis-factor α (TNF-α) and IFN-γ recruiting other effector cells (Wolf, Kissiov, & Raulet, 2023).

The effects of RT on the TME and especially on NK cells appear to depend on various parameters including RT dose, fractionation, target volume and its combination with other immunomodulatory therapies such as immune checkpoint inhibitors (Boustani et al., 2021; Charpentier et al., 2022; Demaria et al., 2021; Deutsch, Chargari, Galluzzi, & Kroemer, 2019). Studying the radio-induced immune response and the role of each parameter is therefore paramount not only to better understand the mechanisms underlying the effects of RT but also to optimize its efficacy.

Flow cytometry is one of the leading techniques to study the TME (McKinnon, 2018). Providing a multi-parametric analysis of single cells, this powerful tool has multiple applications as it can quantify the different cell populations and assess their phenotype as well as the expression of multiple biomarkers (McKinnon, 2018). Thanks to numerous improvements, flow cytometry now offers the possibility of fast low-cost data acquisition and analysis. But its apparent ease of use can lead to errors from the benchtop stages through compensation to analysis (Cossarizza et al., 2021).

Here, we provide a ready-to-use flow cytometry method to characterize the phenotype of infiltrated NK cells within irradiated murine solid tumors.

2. Material

2.1 Equipment

- Humidified cell culture incubator (Memmert, Schwabach, Germany)
- Laboratory safety cabinet (Class II)
- 6-pack Adjustable Volume Pipettes 0.1–2.5, 0.5–10, 2–20, 10–100, 20–200, 100–1000 μL, single channel (Eppendorf, Fisher Scientific, Hamburg, Germany)
- Pipet-Aid (#075212, Drummond®, Dutscher, Brussells, Belgium)
- Optical microscope AE31E (Motic, Hong Kong, China)
- Lab bench centrifuge 460 R (Dutscher)
- Compact anesthesia unit (Minerve, Esternay, France)
- Cordless trimmer for rodent (Bioseblab, Vitrolles, France)
- Conventional cages for rodents (#1284L, Tecniplast, Milan, Italy)
- Animal ventilated cabinet (#BIO-C36 EVO, Tecniplast)
- Digital Caliper: (MOIN1150MI, VWR, Radnor, Pennsylvania, USA)
- Small Animal Radiation Research Platform with motorized variable collimator (X-Strahl, Walsall, United Kingdom)
- GentleMACS™ Dissociator (#130-093-235, Miltenyi, Bergisch-Gladbach, Germany)
- CytoFLEX V5 B5 R3, Research Flow Cytometry (#B53000 Beckman Coulter, Brea, California, USA)
- Stainless steel scissors Wironit 110 mm (#491010, Dutscher)
- Stainless steel dissecting pliers, length 12 cm (#005073, Dutscher)

2.2 Common disposables (see Note 1)

- Sterile 50 mL Falcon® Polypropylene Centrifuge Tubes (#227261, Greiner, Dutscher)
- 0.5 mL Eppendorf tubes (#0030124537, Eppendorf® Fisher Scientific)
- Sterile Serological Pipets (volume 5, 10 or 25 mL, Corning®, Corning, New York, USA)
- Sterile disposable pipette tips (for 1 mL, 200 μL and 10 μL, TipOne® Starlab, Brussels, Belgium)
- Sterile 75 cm^2 tissue-culture treated cell culture flasks, with vented cap (#10430233, Thermo Scientific, Waltham, Massachusetts, USA)
- Hemocytometer for cell counting (Kova® slides, #050126, Dutscher)

- Sterile injection syringes (as an example #070126, capacity 1 mL, Plastipak®, Becton Dickinson, Franklin Lakes, New Jersey, USA)
- Sterile needles 27 G (as an example #300635 Microlance®, Becton Dickinson)
- K-Y® lubricating Jelly Sterile (#250356; Asept InMed, Quint-Fonsegrives, France)
- Single-use scalpel Swann-Morton—N° 15 Stérile (#B01CELEV2K, NMmedical, Mumbai, India)
- GentleMACS™ C-Tubes (#130-096-334, Miltenyi)
- 70 μm cell strainer (#130-110-916, Miltenyi)
- 96-well U-bottom plates (#020028, Dutscher)

2.3 Reagents

- Sterile filtered RPMI 1640 w/L-Glutamine, (#L0500-500, Dutscher)
- Sterile Fetal Bovine Serum (FBS, #500125, Dutscher)
- Sterile Dulbecco's Phosphate Buffered Saline ((PBS) w/o Magnesium, w/o calcium, #L0615-500, Dutscher)
- Sterile filtered 0.5% Trypsin-EDTA (#X0930-100, Dutscher)
- Sterile Corning® trypan blue solution, 0.4% (#702630, Dutscher)
- DNase I (#11284932001, Merck, Rahway, New Jersey, USA)
- Collagenase D (#11088866001, Merck)
- Sterile Dulbecco's Phosphate Buffered Saline (DPBS) w/o Magnesium, w/o calcium, #L0615-500, Dutscher)
- Vetflurane® (Isoflurane, Virbac laboratories, Carros, France)
- FACS Buffer autoMACS® Running Buffer—MACS® Separation Buffer (#130-091-221, Miltenyi)
- DAPI staining solution (#130-111-570, Miltenyi)
- FcR-Block TruStain FcX™ (#101320, Biolegend, San Diego, California, USA)
- CytoFLEX Sheath Fluid (#B51503, Beckman Coulter)
- Fluorochrome-conjugated antibodies (see Table 1)

2.4 Cells and animals

- Colorectal cancer CT26 cell line (American Type Culture Collection)
- Immunocompetent BALB/c mice (Charles River Laboratories, Saint Germain Nuelles, France)

Table 1 Flow cytometry antibody panel.

Reagent	Clone	Fluorochrome	Reference
Anti-CD45	REA 735	PerCP-Vio700	#130-110-801, Miltenyi
Anti-CD3	17A2	A700	#100216, Biolegend
Anti-CD335 (NKp46)	REA815	VioBlue	#130-112-365, Miltenyi
Anti-CD49b	REA981	PE-Vio770	#130-116-434, Miltenyi
Anti-CD11b	M1/70	BV785	#101243, Biolegend
Anti-CD8a	53–6.7	BV605	#100744, Biolegend
Anti-CD314 (NKG2D)	C7	FITC	#115711, Biolegend
Anti-CD159a (NKG2A)	REA1161	PE	#130-120-377, Miltenyi
Anti-KLRG1	REA1016	APC-Vio770	#130-117-093, Miltenyi
Anti-CD184 (CXCR4)	REA107	PE-Vio615	#130-107-610, Miltenyi
Anti-CX3CR1	SA011F11	BV650	#149033, Biolegend
Anti-CD107a	REA777	APC	#130-111-505, Miltenyi
Live and dead cells discrimination	/	Viobility 405/520	#130-130-404, Miltenyi

This panel has been tested for staining using a CytoFlex V5 B5 R3 (Beckman Coulter). CD: cluster of differentiation, NKp46: Natural Killer (NK) protein 46, NKG2D: NK group 2D, NKG2A: NK group 2A, KLRG1: killer cell lectin-like receptor G1, CXCR4: CXC chemokine receptor 4, CX3CR1: CX3C motif chemokine receptor 1.

2.5 Software

- SARRP Control Software, version 4.3.0 (X-Strahl)
- CytExpert V2.5 (Beckman Coulter)

3. Methods

3.1 Cells preparation

1. Expand the murine colorectal cancer CT26 cells in 75 cm^2 flasks in RPMI 1640 medium with 10% FBS in a standard controlled atmosphere (37 °C containing 5% CO_2, 80% humidity) (see Note 2).
2. The day of engraftment, discard the culture medium, wash the adherent cells with 2 mL of pre-warmed PBS.
3. Add 2 mL of 0.5% Trypsin-EDTA solution and incubate for 1–2 min at 37 °C to allow the detachment of CT26 cells from the flask.

4. Add 10 mL of pre-warmed RPMI medium with 10% FBS to the cell suspension for trypsin inactivation (see Note 3). Transfer the cell suspension to a 50 mL tube.
5. Transfer 20 μL of the suspension into a 0.5 mL Eppendorf tube and dilute it with 20 μL of trypan blue (1/2 dilution). Inject 20 μL of the obtained solution into the hemocytometer for cell counting. Estimate the fraction of living cells in the total cell count from at least 3 counting fields (see Note 4).
6. Centrifuge the cell suspension at $300 \times g$ for 5 min at room temperature.
7. Remove the supernatant and resuspend the cell pellet with pre-warmed RPMI medium without FBS to obtain a concentration of 5×10^6 living cells/mL.

3.2 Establishment of CT26 tumors

8. The day before tumor transplantation, shave 8 weeks old BALB/c mice on the right flank to prepare for cancer cell inoculation.
9. On the engraftment day, anesthetize the mice in the induction cage of the Minerve system with 2.5% isoflurane mixed with medical pure oxygen for 2–3 min.
10. Proceed to the injection of 100 μL (*i.e.* 0.5×10^6 cells) of the prepared suspension into the shaved right flank of mice.
11. Ten days after injection, measure the size of tumors with a caliper (see Note 5).
12. Randomize mice with a TV between 125 and 150 mm^3 into the following treatment groups (see Note 6).

3.3 Radiotherapy

13. Once the SARRP is ready to irradiate (see Note 7), anesthetize the mice in the induction cage of the Minerve system with 2.5% isoflurane mixed with medical pure oxygen for 2–3 min.
14. Place mice one after another on the SARRP table with the head introduced in an anesthesia mask delivering 2.5% isoflurane mixed with medical pure oxygen during treatment. In our experiment, the tumor is pinched inside a cup of a home-made device filled with K–Y gel (see Note 8) in order to irradiate the tumor while sparing the mouse.
15. Deliver focal RT to the subcutaneous tumor. In our settings, the SARRP beam is set in therapy mode (220 kV, 13 mA, 0.15 mm copper filter) and the dose is delivered through two opposite beams using a

motorized variable collimator (anterior/posterior, gantry at 0 and 180 °C). In our experiment, the mice receive 3 daily fractions of 8 Gy during 3 following days (see Note 9).

3.4 Flow-cytometry assay

16. At day 3, 7 and 10 after the beginning of radiotherapy, anesthetize the mice in the induction cage of the Minerve system with 2.5% isoflurane mixed with medical pure oxygen for 2–3 min (see Note 10).
17. Sacrifice the anesthetized mice (see Note 11).
18. Collect the subcutaneous tumors avoiding skin into a 24-well plate containing 1 mL of PBS in each well.
19. Transfer the dissected tumors previously cut in small pieces into C-tubes containing 5 mL of a solution of DNase-I (50 μg/mL) and Collagenase-D (1 mg/mL) in serum-free RPMI medium (see Note 12).
20. Insert the C-tubes in a GentleMACS Dissociator and run the following programs in this precise order: m_impTumor_01, 37C_ABDK_1 and m_impTumor_02 (see Note 13).
21. Place the samples on 70-μm filters put on the top of 50 mL tubes and press the samples through using the piston of a syringe.
22. Rinse the C-tubes twice with 10 mL FACS Buffer and poor over the mash trapped within the filters. The final volumes will eventually be the same for each sample.
23. Centrifuge the cell suspension at 300 × *g* at 4 °C for 5 min.
24. Remove the supernatant and resuspend the pellet in 3 mL PBS 1× (see Note 14).
25. In a 96-well U-bottom plate, dilute 10 μL of each sample with 90 μL DAPI staining solution first diluted in PBS (1:100 dilution).
26. Determine the number of living cells through flow cytometry using a DAPI staining (see Notes 15 and 16).
27. In the meantime, centrifuge the 50 mL tubes containing the samples at 300 × *g* at 4 °C for 5 min.
28. After counting, prepare 20×10^6 living cells/mL single-cell suspension in PBS (see Note 17).
29. Transfer 75 μL of each sample into another 96-well U-bottom plate (1.5×10^6 cells per sample).
30. For fluorochromes minus one (FMO), unstained and isotype control wells, mix different samples from different treatment groups. Transfer 75 μL of this mix into 14 wells (1 well per FMO, 1 well for the isotype control and 1 well for the unstained control) (see Note 18).

31. Centrifuge the cell suspension at 300 × *g* at 4 °C for 5 min.
32. Discard the supernatant, add 100 μL Viobility diluted in PBS (1:100 dilution) to each well but the unstained sample one and incubate at room temperature for 15 min in darkness (see Note 15).
33. Centrifuge the cell suspension at 300 × *g* at 4 °C for 5 min.
34. Discard the supernatant and resuspend the pellet in 200 μL FACS Buffer.
35. Centrifuge the cell suspension at 300 × *g* at 4 °C for 5 min.
36. Discard the supernatant again and add 50 μL mouse FcBlock to each well and incubate at 4 °C for 5 min (see Note 19).
37. For cellular staining, add 50 μL antibody mix (see Table 1) diluted in FACS Buffer to each sample well, 50 μL of each FMO mix to the corresponding wells, 50 μL of the isotype mix to the isotype well and 50 μL FACS Buffer to the unstained well (see Note 20). Incubate at 4 °C for 20 min in darkness (final volume per well 100 μL).
38. Centrifuge the cell suspension at 300 × *g* at 4 °C for 5 min. Resuspend the pellet in 200 μL FACS Buffer. Perform this step twice.
39. Insert the 96-well plate in the cytometer and acquire. To avoid clogging the device and limit doublets, acquire at slow or medium speed depending on the event rate. Regularly check that the time gate is constant during acquisition to detect possible clogging events.

3.5 FACS analysis

The compensation process correcting for fluorescence spillover is not discussed here as it is not specific to our NK cell characterization. Cossarizza et al. (2021) have published guidelines on this subject the readers can refer to. To further analyze the different populations within the samples, it is important to first identify them through a proper gating strategy. This is detailed below and illustrated in Fig. 1.

40. Draw a first gate selecting cells depending on their size (forward scatter area, FSC-A) and granularity (side scatter area, SSC-A). Exclude from the gate the debris found close to the graph origin.
41. To get rid of doublets, use a graph with FSC-A and FSC-H (FSC height) and gate out all events that appear out of the diagonal that corresponds to single-cell events.
42. Within single cells, remove dead cells by selecting only negative events for Viobility (see Note 15).
43. Identify immune cells by selecting $CD45^+$ cells. Within $CD45^+$ cells, isolate lymphocytes that are small in size according to SSC-A.

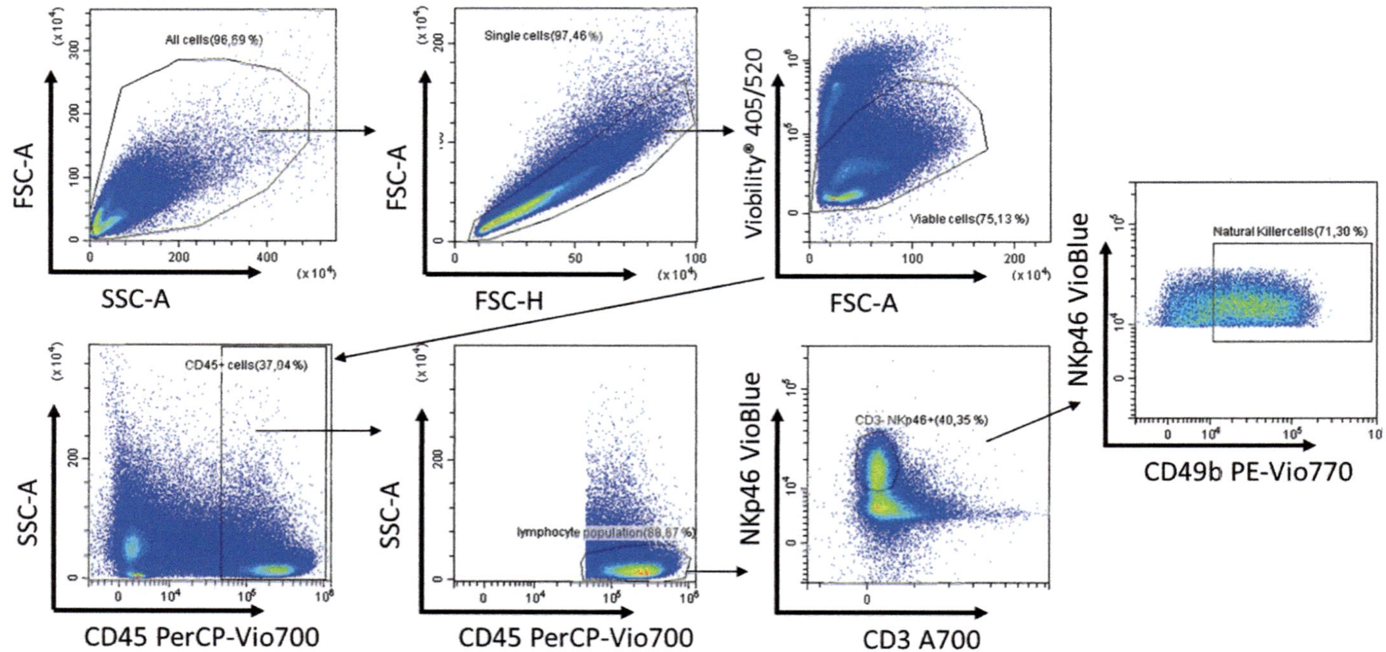

Fig. 1 Gating strategy to characterize Natural Killer cells infiltrating irradiated CT26 tumors in BALB/c mice. Proposition of gating strategy that can be used to select NK cells infiltrated within the tumor to further characterize them according to the antibody panel described in Table 1. NK cells are identified as viable CD45+ CD3− NKp46+ CD49b+ cells. FSC-A/H: forward scatter area/height, SSC-A: side scatter area, CD: cluster of differentiation, NKp46: Natural Killer protein 46.

The exclusion of non-lymphocytic populations helps getting rid of non-specific NKp46 staining on subsequent NK cell gating.

44. Within lymphocytes, identify cells positive for NKp46 and negative for CD3. Inside this NKp46$^+$ CD3$^-$ population, identify NK cells as positive events for CD49b (Note 21).

45. Different NK cell populations can eventually be observed through the expression of the different biomarkers from the panel. Gating is facilitated by the use of FMO for each antibody, especially for those for which the discrimination between positive and negative populations is not apparent (Note 22).

4. NK cell characterization

NK cells are an actively studied lymphoid component whose phenotypic characterization does not always meet with consensus, since certain markers do not provide the same information depending on the team involved (Abel, Yang, Thakar, & Malarkannan, 2018; Judge, Murphy, & Canter, 2020; Ma, Caligiuri, & Yu, 2022; Vivier et al., 2011). For instance, NKp46 is considered as a conserved marker that best defines NK cells for some authors (Vivier et al., 2011), but as a protein acquired at late maturation stages for others (Ma et al., 2022). In parallel, Killer cell lectin-like receptor G1 (KLRG1) is seen as an end stage maturation markers by some teams (Abel et al., 2018) and is used to define the NK cell exhaustion phenotype for others (Judge et al., 2020).

We propose here a flow cytometry antibody panel to analyze NK cell maturation, activation, cytotoxicity and chemotaxis (see Table 1). In BALB/c mice, NK cells do not express NK1.1 and can be identified through the expression of NKp46 and CD49b (Cossarizza et al., 2021), and were hence defined in our analysis as CD45$^+$ CD3$^-$ NKp46$^+$ CD49b$^+$ cells.

Briefly, CD11b is a marker progressively acquired by NK cells during maturation (Abel et al., 2018; Ma et al., 2022). NK group 2D (NKG2D) and NK group 2A (NKG2A) are activating and inhibitory receptors, respectively (Pegram et al., 2011). Killer cell lectin-like receptor G1 (KLRG1) is, as discussed, associated with end-stage maturation or exhaustion (Abel et al., 2018; Judge et al., 2020). CXC chemokine receptor 4 (CXCR4) and CX3C motif chemokine receptor 1 (CX3CR1) are receptors to CXC motif chemokine ligand 12 (CXCL12) and CX3CL1, respectively, and involved in the chemotaxis and migration of NK cells (Ran et al., 2022). At last, CD107a is expressed by cytotoxic cells that have degranulated, and is therefore an indirect marker of cytotoxicity (Aktas, Kucuksezer, Bilgic, Erten, & Deniz, 2009).

The growing knowledge of NK cells will probably lead to a better understanding of their biology. In the meantime, the creation of a cytometry panel designed for NK cell analysis will depend on the purpose of the experiments carried out.

5. Notes

Note 1. All references in the disposable product catalogue can be replaced by equivalent products purchased from different suppliers.

Note 2. Mycoplasma contamination of cell cultures is a concern, especially when studying the immune infiltrate. We therefore recommend regular mycoplasma testing.

Note 3. FBS is key at this step as it is serum that inactivates trypsine.

Note 4. Trypan blue is used to discriminate live and dead cells as it enters within unviable cells presenting membrane damage.

Note 5. Tumor volumes are estimated according to the following formula: $TV = 0.5 \times length \times width^2$.

Note 6. In our experience, a TV of 125–150 mm^3 is reached at day 10 after tumor transplantation with CT26 tumors. The required TV and the time needed to reach it depends on tumor models.

Note 7. Our radiation treatment planification and procedure is extensively detailed elsewhere (Mirjolet, Ladjohounlou, Baude, & Bessières, 2023, In Press). Other devices can be used to deliver RT, such as the X-RAD SmART (X-Ray Inc., USA).

Note 8. This home-made restraint device was designed to perform routinely murine subcutaneous tumors and is also described elsewhere (Mirjolet et al., 2023, In Press).

Note 9. Depending on the purpose of the experiment, several doses and/or fractionations can be delivered. When comparing different fractionations, make sure that your regimens present an equal biologically effective dose, *i.e.* the total absorbed dose by biological tissues.

Note 10. In immunology and radiobiology, timing is key, as RT immunomodulates the TME in a time-dependent manner. As a consequence, it is crucial to determine precise time points to evaluate the role of RT. To characterize infiltrated NK cells, we have found that early time points such as day 3, 7 and 10 are relevant.

Note 11. In our team, mice are sacrificed through cervical dislocation, as this technique has been previously approved for our experiments by an independent ethics committee and by the Société du Bien-Être Animal (SBEA, *society for the welfare of animals*).

Note 12. DNase-I and Collagenase-D are essential for a good quality flow cytometry experiment. DNase-I helps prevent cell clumping from DNA released from lysed cells during tissue digestion, whereas Collagenase-D degrades collagen present in the extracellular matrix.

Note 13. Even though it is not absolutely necessary, the use of a tissue dissociator such as a Gentle MACS Dissociator mechanically helps tissue digestion. Other programs can be run depending on the tumor models.

Note 14. The lysis of red blood cells (RBC) is an option at this stage. However, as NK cells constitute a relatively small tumor-infiltrating lymphocyte population, the number of events for each marker may be low, and lysing RBC may be responsible for the loss of some NK cells and potentially compromise the analysis due to a lack of events.

Note 15. The cell count can be manually performed by following **step 5**.

Note 16. Viability dyes such as DAPI or Viobility® only penetrate into cells presenting membrane damages.

Note 17. This cell concentration is sufficient to obtain enough events for the characterization of NK cells in our model. However, depending on the tumor model and the markers to analyze, this number may vary.

Note 18. FMO, unstained and isotype controls are key to good quality flow cytometry analysis. FMO controls are samples stained with all the fluorochromes of the panel minus one that are used to help properly gate populations in multicolor experiments such as this one. Unstained control is used to determine the background autofluorescence of cells. The isotype control is a sample in which are mixed antibodies that differ from the one from the panel only for the specificity of the antigen. It is important to evaluate the specificity of the primary antibodies, enabling the differentiation between specific and non-specific staining signal caused by off-target interactions with cell surfaces.

Note 19. Blocking Fc receptors before staining is an important step to limit non-specific off-target interactions between cells and antibodies through their Fc domains.

Note 20. Antibody concentrations are not provided here, as the volume of each antibody in the different mixes has to be determined prior to the experiment by antibody titration. Titration implies performing antibody

dilutions and staining the same number of cells in the same volume. The dilution to be used is the one representing the best stain index.
Note 21. It is important to notice that mouse strains such as BALB/c mice do not react with anti-NK1.1 antibodies. Therefore, we use anti-NKp46 and anti-CD49b antibodies to detect NK cells in our panel.
Note 22. It is recommended to include FMOs for each antibody, but FMOs appear to be particularly useful for markers for which positive and negative populations are difficult to separate.

6. Conclusion

With immunology now inextricably linked to radiobiology, flow cytometry has become an important tool for radiobiology teams. Good practices are key to high-quality analysis leading to robust results. We have presented a protocol to better characterize NK cells within irradiated tumors, as it currently represents a hot topic. Before starting the cytometry experiment, many parameters likely to modify the results need to be considered, such as the timing, the antibody panel designed and the model tumor. Apart from a few minor adjustments, the method described here may help radiobiologist wishing to study RT and NK cells through flow cytometry.

Acknowledgments

C.M. team was supported for this research topic by "CCIR-GE" (Ligue Régionale contre le Cancer Comité Grand Est), by "Conseil Régional Bourgogne Franche-Comté" and by CGFL (Centre Georges-François Leclerc). J.B. is supported by the ARC Fondation (Association de recherche contre le Cancer).

Declaration of interests

The authors declare no conflict of interest.

References

Abel, A. M., Yang, C., Thakar, M. S., & Malarkannan, S. (2018). Natural killer cells: Development, maturation, and clinical utilization. *Frontiers in Immunology*, *9*, 1869. https://doi.org/10.3389/fimmu.2018.01869.

Aktas, E., Kucuksezer, U. C., Bilgic, S., Erten, G., & Deniz, G. (2009). Relationship between CD107a expression and cytotoxic activity. *Cellular Immunology*, *254*, 149–154. https://doi.org/10.1016/j.cellimm.2008.08.007.

Bardhan, K., Anagnostou, T., & Boussiotis, V. (2016). The PD1:PD-L1/2 pathway from discovery to clinical implementation. *Frontiers in Immunology*. https://doi.org/10.3389/fimmu.2016.00550.

Baude, J., Limagne, E., Ladjohounlou, R., & Mirjolet, C. (2023). Combining radiotherapy and NK cell-based therapies: The time has come. In *International review of cell and molecular biology* Academic Press. https://doi.org/10.1016/bs.ircmb.2023.02.003.

Bernier, J., Hall, E., & Giaccia, A. (2004). Radiation oncology: A century of achievements. *Nature Reviews. Cancer*. https://doi.org/10.1038/nrc1451.

Boustani, J., Grapin, M., Laurent, P.-A., Apetoh, L., & Mirjolet, C. (2019). The 6th R of radiobiology: Reactivation of anti-tumor immune response. *Cancers (Basel), 11*. https://doi.org/10.3390/cancers11060860.

Boustani, J., Lecoester, B., Baude, J., Latour, C., Adotevi, O., Mirjolet, C., et al. (2021). Anti-PD-1/anti-PD-L1 drugs and radiation therapy: Combinations and optimization strategies. *Cancers (Basel), 13*. https://doi.org/10.3390/cancers13194893.

Canter, R., Grossenbacher, S., Foltz, J., Sturgill, I., Park, J., Luna, J., et al. (2017). Radiotherapy enhances natural killer cell cytotoxicity and localization in pre-clinical canine sarcomas and first-in-dog clinical trial. *Journal for Immunotherapy of Cancer*. https://doi.org/10.1186/s40425-017-0305-7.

Charpentier, M., Spada, S., Van Nest, S. J., & Demaria, S. (2022). Radiation therapy-induced remodeling of the tumor immune microenvironment. *Seminars in Cancer Biology, 86*, 737–747. https://doi.org/10.1016/j.semcancer.2022.04.003.

Chen, D., Patel, R., Verma, V., Ramapriyan, R., Barsoumian, H., Cortez, M., et al. (2020). Interaction between lymphopenia, radiotherapy technique, dosimetry, and survival outcomes in lung cancer patients receiving combined immunotherapy and radiotherapy. *Radiotherapy and Oncology: Journal of the European Society for Therapeutic Radiology and Oncology*. https://doi.org/10.1016/j.radonc.2020.05.051.

Chiossone, L., Dumas, P., Vienne, M., & Vivier, E. (2018). Natural killer cells and other innate lymphoid cells in cancer. *Nature Reviews. Immunology*. https://doi.org/10.1038/s41577-018-0061-z.

Cossarizza, A., Chang, H. D., Radbruch, A., Abrignani, S., Addo, R., Akdis, M., et al. (2021). Guidelines for the use of flow cytometry and cell sorting in immunological studies (third edition). *European Journal of Immunology*. https://doi.org/10.1002/eji.202170126.

De Martino, M., Daviaud, C., & Vanpouille-Box, C. (2021). Radiotherapy: An immune response modifier for immuno-oncology. *Seminars in Immunology*. https://doi.org/10.1016/j.smim.2021.101474.

Demaria, S., Guha, C., Schoenfeld, J., Morris, Z., Monjazeb, A., Sikora, A., et al. (2021). Radiation dose and fraction in immunotherapy: One-size regimen does not fit all settings, so how does one choose? *Journal for Immunotherapy of Cancer*. https://doi.org/10.1136/jitc-2020-002038.

Deutsch, E., Chargari, C., Galluzzi, L., & Kroemer, G. (2019). Optimising efficacy and reducing toxicity of anticancer radioimmunotherapy. *The Lancet Oncology, 20*, e452–e463. https://doi.org/10.1016/S1470-2045(19)30171-8.

Dovedi, S., & Illidge, T. (2015). The antitumor immune response generated by fractionated radiation therapy may be limited by tumor cell adaptive resistance and can be circumvented by PD-L1 blockade. *Oncoimmunology*. https://doi.org/10.1080/2162402X.2015.1016709.

Gao, R., Shi, G., & Wang, J. (2022). Functional diversities of regulatory T cells in the context of Cancer immunotherapy. *Frontiers in Immunology*. https://doi.org/10.3389/fimmu.2022.833667.

Grapin, M., Richard, C., Limagne, E., Boidot, R., Morgand, V., Bertaut, A., et al. (2019). Optimized fractionated radiotherapy with anti-PD-L1 and anti-TIGIT: A promising new combination. *Journal for Immunotherapy of Cancer*, 7, 160. https://doi.org/10.1186/s40425-019-0634-9.

Judge, S., Murphy, W., & Canter, R. (2020). Characterizing the dysfunctional NK cell: Assessing the clinical relevance of exhaustion, anergy, and senescence. *Frontiers in Cellular and Infection Microbiology*. https://doi.org/10.3389/fcimb.2020.00049.

Kang, C., Jeong, S.-Y., Song, S. Y., & Choi, E. K. (2020). The emerging role of myeloid-derived suppressor cells in radiotherapy. *Radiation Oncology Journal*, *38*, 1–10. https://doi.org/10.3857/roj.2019.00640.

Lhuillier, C., Rudqvist, N.-P., Elemento, O., Formenti, S. C., & Demaria, S. (2019). Radiation therapy and anti-tumor immunity: Exposing immunogenic mutations to the immune system. *Genome Medicine*, *11*, 40. https://doi.org/10.1186/s13073-019-0653-7.

Liu, S., Sun, X., Luo, J., Zhu, H., Yang, X., Guo, Q., et al. (2015). Effects of radiation on T regulatory cells in normal states and cancer: Mechanisms and clinical implications. *American Journal of Cancer Research*, *5*, 3276–3285.

Lo Nigro, C., Macagno, M., Sangiolo, D., Bertolaccini, L., Aglietta, M., & Merlano, M. C. (2019). NK-mediated antibody-dependent cell-mediated cytotoxicity in solid tumors: Biological evidence and clinical perspectives. *Annals of Translational Medicine*, 7, 105. https://doi.org/10.21037/atm.2019.01.42.

Ma, S., Caligiuri, M., & Yu, J. (2022). A four-stage model for murine natural killer cell development in vivo. *Journal of Hematology & Oncology*. https://doi.org/10.1186/s13045-022-01243-1.

McKinnon, K. (2018). Flow cytometry: An overview. *Current Protocols in Immunology*. https://doi.org/10.1002/cpim.40.

McLaughlin, M., Patin, E., Pedersen, M., Wilkins, A., Dillon, M., Melcher, A., et al. (2020). Inflammatory microenvironment remodelling by tumour cells after radiotherapy. *Nature Reviews. Cancer*. https://doi.org/10.1038/s41568-020-0246-1.

Mirjolet, C., Ladjohounlou, R., Baude, J., & Bessières, I. (2023). Novel platform for sub-cutaneous tumor irradiation in mice. *Methods in Cell Biology*. https://doi.org/10.1016/bs.mcb.2023.06.004.

Pegram, H. J., Andrews, D. M., Smyth, M. J., Darcy, P. K., & Kershaw, M. H. (2011). Activating and inhibitory receptors of natural killer cells. *Immunology and Cell Biology*, *89*, 216–224. https://doi.org/10.1038/icb.2010.78.

Punjabi, A., Barrett, E., Cheng, A., Mulla, A., Walls, G., Johnston, D., et al. (2021). Neutrophil-lymphocyte ratio and absolute lymphocyte count as prognostic markers in patients treated with curative-intent radiotherapy for non-small cell lung cancer. *Clinical Oncology (Royal College of Radiologists (Great Britain))*. https://doi.org/10.1016/j.clon.2021.03.019.

Ran, G., Lin, Y., Tian, L., Zhang, T., Yan, D., Yu, J., et al. (2022). Natural killer cell homing and trafficking in tissues and tumors: From biology to application. *Signal Transduction and Targeted Therapy*. https://doi.org/10.1038/s41392-022-01058-z.

Shevtsov, M., Sato, H., Multhoff, G., & Shibata, A. (2019). Novel approaches to improve the efficacy of immuno-radiotherapy. *Frontiers in Oncology*. https://doi.org/10.3389/fonc.2019.00156.

Tanchot, C., Terme, M., Pere, H., Tran, T., Benhamouda, N., Strioga, M., et al. (2013). Tumor-infiltrating regulatory T cells: Phenotype, role, mechanism of expansion in situ and clinical significance. *Cancer Microenvironment: Official Journal of the International Cancer Microenvironment Society*. https://doi.org/10.1007/s12307-012-0122-y.

Vanpouille-Box, C., Diamond, J., Pilones, K., Zavadil, J., Babb, J., Formenti, S., et al. (2015). TGFβ is a master regulator of radiation therapy-induced antitumor immunity. *Cancer Research*. https://doi.org/10.1158/0008-5472.CAN-14-3511.

Vivier, E., Raulet, D. H., Moretta, A., Caligiuri, M. A., Zitvogel, L., Lanier, L. L., et al. (2011). Innate or adaptive immunity? The example of natural killer cells. *Science*, *331*, 44–49. https://doi.org/10.1126/science.1198687.

Vivier, E., Tomasello, E., Baratin, M., Walzer, T., & Ugolini, S. (2008). Functions of natural killer cells. *Nature Immunology*, *9*, 503–510. https://doi.org/10.1038/ni1582.

Walle, T., Martinez Monge, R., Cerwenka, A., Ajona, D., Melero, I., & Lecanda, F. (2018). Radiation effects on antitumor immune responses: Current perspectives and challenges. *Therapeutic Advances in Medical Oncology*, *10*. https://doi.org/10.1177/1758834017742575.

Wang, N., Lei, Z., Yang, H., Tang, Z., Yang, M., Wang, Y., et al. (2022). Radiation-induced PD-L1 expression in tumor and its microenvironment facilitates cancer-immune escape: A narrative review. *Annals of Translational Medicine*. https://doi.org/10.21037/atm-22-6049.

Wolf, N. K., Kissiov, D. U., & Raulet, D. H. (2023). Roles of natural killer cells in immunity to cancer, and applications to immunotherapy. *Nature Reviews. Immunology*, *23*, 90–105. https://doi.org/10.1038/s41577-022-00732-1.

Yi, M., Zheng, X., Niu, M., Zhu, S., Ge, H., & Wu, K. (2022). Combination strategies with PD-1/PD-L1 blockade: Current advances and future directions. *Molecular Cancer*. https://doi.org/10.1186/s12943-021-01489-2.

CHAPTER NINE

Flow cytometry detection and quantification of circulating leukocyte subpopulations in mice after brain irradiation

Julie Coupey[a], Marine M. Leblond[b,c], Erika S. Hue[d,e], and Samuel Valable[a,*]

[a]Université de Caen Normandie, CNRS, Normandie Université, ISTCT UMR6030, GIP Cyceron, Caen, France
[b]Department of Oncology UNIL CHUV, University of Lausanne, Lausanne, Switzerland
[c]Ludwig Institute for Cancer Research, University of Lausanne, Lausanne, Switzerland
[d]LABÉO, Saint-Contest, France
[e]Normandie Université, UNICAEN BIOTARGEN, Saint-Contest, France
[*]Corresponding author: e-mail address: samuel.valable@cnrs.fr

Contents

Methods in Cell Biology, Volume 189
ISSN 0091-679X
https://doi.org/10.1016/bs.mcb.2024.06.004

Abstract

In the context of high-grade gliomas such as glioblastoma (GBM), the immune part of the tumor microenvironment (TME) is involved in tumor growth and tumor recurrence. It is mostly represented by high amount of macrophages and low amount of lymphocytes. GBM in itself as well as x-ray-based radiotherapy, a standard treatment for brain tumors, are also associated with systemic effects like lymphopenia that correlates with a poor prognosis. This contributes to the immune-suppressive nature of the TME and may explain the lack of the anti-tumor immune response. Radiation-induced lymphopenia (RIL) is generally evaluated on $CD4^+$ and $CD8^+$ count or on a CBC (complete blood count), but the heterogeneity of the subtypes prompts us to explore them in detail to better understand the cellular response to brain irradiation.

To facilitate and develop the evaluation of x-ray brain exposure on circulating immune cells, we developed a reproducible and reliable method to quantify the variation of lymphoid and myeloid subtypes using flow cytometry after brain irradiation in the rodent.

1. Introduction

Conventional x-rays-based radiotherapy remains one of the most used therapeutic option for supratentorial brain tumors. Tumor cells are the main target of the treatment but beyond them, the complexity of the tumor microenvironment (TME) and the importance of the crosstalk between cancer cells and the TME is being increasingly established to limit treatment efficacy (Tomaszewski, Sanchez-Perez, Gajewski, & Sampson, 2019).

In the context of high-grade brain tumors such as GBM, the TME can be considered as a multicompartmental system among which hypoxia, angiogenesis and the immune system (Jarosz-Biej, Smolarczyk, Cichoń, & Kułach, 2019; Quail & Joyce, 2017) are the main players (Leblond et al., 2016; Collet et al., 2021). In recent years, study of the immune system has gained a great interest in the cancer field with recent observations that immune cells are involved in the control of tumor growth (Alifieris & Trafalis, 2015; Wang & Haffty, 2018). The immune part of the TME is mainly composed of lymphoid and myeloid populations. In the field of brain neoplasms, tumor-associated macrophages (TAMs) are the major immune population, coming from microglia and circulating monocytes (Klemm et al., 2020). Despite of their origin, different subpopulation of TAMs from anti-tumor/pro-inflammatory to pro-tumor/anti-inflammatory TAMs can be present simultaneously in high-grade brain tumor. Anti-inflammatory TAMs are preferentially found within the tumor and their elevated number is associated with high grade and poor prognosis in patients (Prosniak et al., 2013).

On the contrary, tumor-infiltrating lymphocytes play a substantial role in tumor regression and response to treatment. However, lymphocytes are poorly present in GBM (Klemm et al., 2020). This could be explained by the fact that the tumor itself induces lymphopenia (<200/μL vs 1000/μL in treatment-naïve control patients) in treatment-naïve GBM patients (Chongsathidkiet et al., 2018) that could limit the number of circulating T cells to enter the tumor site. The low lymphocyte number in the brain could also be the result of anti-tumor treatments on these leukocyte populations. Add to this the fact that the balance between the proportion of $CD4^+$ and $CD8^+$ lymphocytes is associated with a poor prognosis and tumor regression in GBM (Han et al., 2014); this could also explain the difficulties in treating high-grade brain tumors with immunotherapy.

In addition to the tumor context, the investigation of ionizing radiations' direct action on the biological function of cells in the TME is of great interest. Indeed, the radiation deposited dose is a factor influencing the phenotype and function of immune cells in the TME. Our team previously reported that conventional x-rays favored pro-tumor TAMs in the tumor (Leblond et al., 2017). Literature also demonstrated that it can induced the recruitment of myeloid cells into the tumor that may be responsible for tumor recurrence (Akkari et al., 2020; Rafat et al., 2018). Another known adverse effect of ionizing radiation is the reduction of lymphocytes in the blood, or radiation-induced lymphopenia (RIL) in several forms of cancer, including central nervous system (CNS) tumors (Campian, Sarai, Ye, Marur, & Grossman, 2014; Ghosh et al., 2023; Huang et al., 2015; Pham et al., 2024). RIL can even occur after irradiation of low containing lymphoid organs in the radiation zone, such as brain irradiation. T cells are indeed the most radiosensitive immune cells in the body, with a D10 (dose required to have a survival fraction equal to 10%) estimated at 3 Gy (Nakamura, Kusunoki, & Akiyama, 1990; Pham et al., 2023). In a context of GBM, Grossman et al. showed that for more than 10% of patients, the $CD4^+$ T cells count was below $200/mm^3$ ($500/mm^3 < \text{baseline} < 1500/mm^3$) after the end of chemo-radiotherapy (Grossman et al., 2011). Furthermore, in a modeling irradiation study, Yovino et al. demonstrated that after 30 fractions of radiotherapy, 99% of circulating cells received a dose greater than 0.5 Gy (Yovino, Kleinberg, Grossman, Narayanan, & Ford, 2013). Since, studies have shown that this RIL can persist for several months and can lead to reduced survival as well as accelerated tumor progression (Cesaire, Le Mauff, Rambeau, Toutirais, & Thariat, 2020; Dutoit et al., 2020; Grossman et al., 2015). In addition, the incidence and severity of RIL may be dependent on dosimetric factors, such as the duration of treatment, the number of fractions and the ionizing radiation type (Ahn et al., 2020; Grossman et al., 2011; Hammi, Paganetti, & Grassberger, 2020; Kumar et al., 2020; Macintyre et al., 2021; Mohan et al., 2020). This effect of conventional x-rays may explain the lack of the anti-tumor immune response in the treatment of high-grade brain tumors. RIL is generally evaluated on $CD4^+$ and $CD8^+$ count or on a CBC (complete blood count), but the heterogeneity of the subtypes prompts us to explore them in details.

To facilitate and develop the evaluation of x-ray brain exposure on circulating immune cells, we developed a home-made method to evaluate and quantify the variation of lymphoid and myeloid lineages after brain irradiation in a reproducible and reliable way in mouse. From whole peripheral blood, alive hematopoietic cells ($dead^-CD45^+$) were selected to gate on leukocyte's lineages: myeloid ($CD45^+CD11b^+$), T ($CD45^+CD3^+$), B ($CD45^+CD3^-B220^+$) and NK ($CD45^+CD3^-NK1.1^+$) cells allowing for the subpopulation dosage to be refined. From myeloid cells, monocytes ($CD11b^+SiglecF^-Ly6C^+$), neutrophils ($CD11b^+SiglecF^-Ly6G^+$) and eosinophils ($CD11b^+SiglecF^+$) were selected. From T lymphocytes, CD8 T cells ($CD3^+CD8^+$) and conventional CD4 T cells ($CD3^+CD4^+FoxP3^-$) were selected. A gating strategy was then developed to identify CD8 and CD4 T lymphocyte subtypes in greater detail.

Here we describe a single-cell analysis method to evaluate circulating inflammation by measuring circulating leukocyte subpopulations after brain irradiation. Using flow cytometry, we followed in time the evolution of myeloid and lymphoid subpopulation circulating cells after x-ray brain exposure in immunocompetent tumor-free C57BL/6 mice.

2. Material

2.1 Disposables

- 4 mm tip sterile single-use lancet (*GR-4MM*, Goldenrod animal lancet, BiosebLab)
- 96-wells V-bottom microplate (#*650101*, Greiner Bio-one)
- Pre-heparinized microtube such as PP micro test tube, 1.5 mL, blue Hep-Li coated, with stopper Φ10.9 × 39.8 mm (#*041-245,510*, Dutscher)
- 15 mL Falcon (as an example, #*2583004*, Sarstedt)
- 50 mL Falcon (as an example, #*2583167*, Sarstedt)
- 5 L polystyrene round-bottom tube (#*352054*, Falcon®)
- Multichannel Pipette
- Monochannel Pipettes

- Pipette tips (aspirating 1 mL, 200 µL and 10 µL)
- Pipetting reservoirs
- Electronic scales for ultra-precise weighing

Note: the disposable references can be replaced by equivalent references.

2.2 Mice and reagents

- Female C57BL/6 mice aged 6–7 weeks on day of experiment (16–20 g, Janvier Labs Laboratories, France)
- Ultra-pure water
- Hydrating jelly (DietGel Recovery, #*72-06-5022*, SSNIFF Spezialdiäten GmbH)
- 1× Dulbecco's phosphate buffered saline (#*CS1PBS01-01*, Eurobio Scientific), PBS
- Sodium azide ReagentPlus®, ≥99.5% (#*S2002*, Sigma Aldrich)
- Ethylenediaminetetraacetic acid ACS reagent (#*798681*, Sigma Aldrich), EDTA
- Bovine Serum Albumin (#*A9418*, Sigma Aldrich), BSA
- Anesthetic gas for inhalation in vial (Isoflurane Belamont)
- 1× red blood cell lysis buffer (#*00433357*, Thermo Fisher)
- Live/dead™ Fixable Aqua Dead Cell Stain Kit *(#L34966*, Thermo Fisher)
- Fixation/Permeabilization Concentrate and Diluent *(#00512343*, #*00522356*, Thermo Fisher)
- 10× permeabilization buffer (#*00833356*, Thermo Fisher)
- Antibodies listing (Table 1)
- Counting beads (#*11570066*, Thermo Fisher)
- UltraComp eBeads™ (#*0122242*, Thermo Fisher)

2.3 Reagent preparation

- FACS buffer: prepare a fluorescent activated cell sorting buffer in PBS 1× with 2.92% EDTA, 0.2% BSA and 0.2% sodium azide. The final solution needs to be stored at 4 °C when not in use.

2.4 Equipment

- High-speed centrifuge (5430R, Eppendorf)
- X-ray irradiator such as the small animal irradiator XRad-225Cx (Precision XRay Inc.)
- Flow cytometer equipped with 4 lasers/15 detectors for acquisition of fluorescence such as the Cytoflex S® (Beckman Coulter)

Table 1 Commercially available flow cytometry antibodies.

Target	Cellular subtype	Fluorochrome	Reference	Supplier	Concentration
CD16/CD32	Fcγ receptor	–	101,320	Biolegend™	0.5 mg/mL
LIVE/DEAD™ Fixable Aqua Dead Cell Stain Kit	Dead cell	526 nm emission	L34966	Invitrogen	–
CD45.2	Hematopoietic cell	BV650	109,836	Biolegend™	0.2 mg/mL
CD3	T lymphocyte	BV605	100,237	Biolegend™	100 μg/mL
CD4	Helper T cell	PE	12,004,282	Invitrogen	0.2 mg/mL
CD8	Cytotoxic T cell	PE/Cy7	100,722	Biolegend™	0.2 mg/mL
B220 (CD45.R)	B lymphocyte	Percp/Cy5.5	103,236	Biolegend™	0.2 mg/mL
NK1.1	Natural killer	FITC	11,594,182	Invitrogen	0.5 mg/mL
RORγt	Helper T cell 17 (Th17)	PE/TR	61,698,182	Invitrogen	0.2 mg/mL
FoxP3	Regulatory T cell	BV421	48,577,382	Invitrogen	0.2 mg/mL
CD44	Effector T cell	FITC	11,044,182	Invitrogen	0.5 mg/mL
CD62L	Effector memory T cell	Percp/Cy5.5	45,062,182	Invitrogen	0.2 mg/mL
PD-1 (CD279)	T cell activation Exhaustion marker	PE	135,206	Biolegend™	0.2 mg/mL

CTLA4 (CD152)	T cell activation Exhaustion marker	BV421	106,312	Biolegend™	0.2 mg/mL
CD11b	Myeloid cell	Percp/Cy5.5	550,993	BD Biosciences™	0.2 mg/mL
SiglecF	Eosinophil	PE/TR	130,112,172	Miltenyi Biotec™	1.5 μg/mL
Ly6C	Monocyte	BV605	563,011	BD Biosciences™	0.2 mg/mL
Ly6G	Neutrophil	BV421	127,628	Biolegend™	0.2 mg/mL
MHCII	Monocyte subtype	PE/Cy7	107,630	Biolegend™	0.2 mg/mL
CX3CR1	Monocyte subtype	PE	149,006	Biolegend™	0.2 mg/mL
F4/80	Macrophage	FITC	11,480,182	Invitrogen	0.5 mg/mL

2.5 Software

- SmARTPlan TPS (Precision XRay Inc.)
- 5-Pilot XRad 225 pilot 1.10.6 software (Precision XRay Inc.)
- CytExpert 2.3.1.22 software (Beckman Coulter)
- Kaluza 2.1 (Beckman Coulter)
- GraphPad Prism 9.0 (San Diego, CA)
- Excel 2019 (Microsoft)

3. Method

All in vivo radiation experiments were performed on the small animal irradiator XRad-225Cx (Cyceron platform, Caen, France) and in accordance with the current European regulations, with permission of the regional committees on animal ethics CENOMEXA (#27343).

3.1 Radiation treatment plan preparation

Treatment planning was performed using the SmARTPlan Treatment planning software (TPS) for small animal radiotherapy studies. The x-ray beam can be parameterized to adjust the dose and rate of irradiation (Fig. 1A–E).

3.2 Radiation treatment procedure

Mice were placed in a cradle (Minerve) and were irradiated with a 20 Gy dose under anesthesia (1 L/min with 5% isoflurane for induction and 2% for maintenance in 70% N_2O/30% O_2). An 8 mm square collimator was used for whole-brain and left hemisphere irradiations. The same collimator is used for both irradiation conditions. First, a scanner control with an aluminum filter (80 keV/0.5 mA) is performed to set the beam. The image obtained enables the beam to be adjusted according to the targeted irradiation zone (in this case, whole-brain (Fig. 1A) or left hemisphere (Fig. 1B)) and to avoid high-risk areas (in this case, the olfactory bulbs and brainstem (Fig. 1C)). Then, the treatment was applied using a copper filter (225 keV/13 mA) (Fig. 1D and E).

3.3 Sample collection

The blood was collected submandibularly as described in the literature and as recommended in practice guidelines (Francisco, Howarth, & Whittaker, 2015; Golde, Gollobin, & Rodriguez, 2005; Jo, Bae, Yoon, Kim, & Han, 2021). Indeed, submandibular bleeding is an easy way to collect 0.05–0.3 mL of blood without the use of anesthesia.

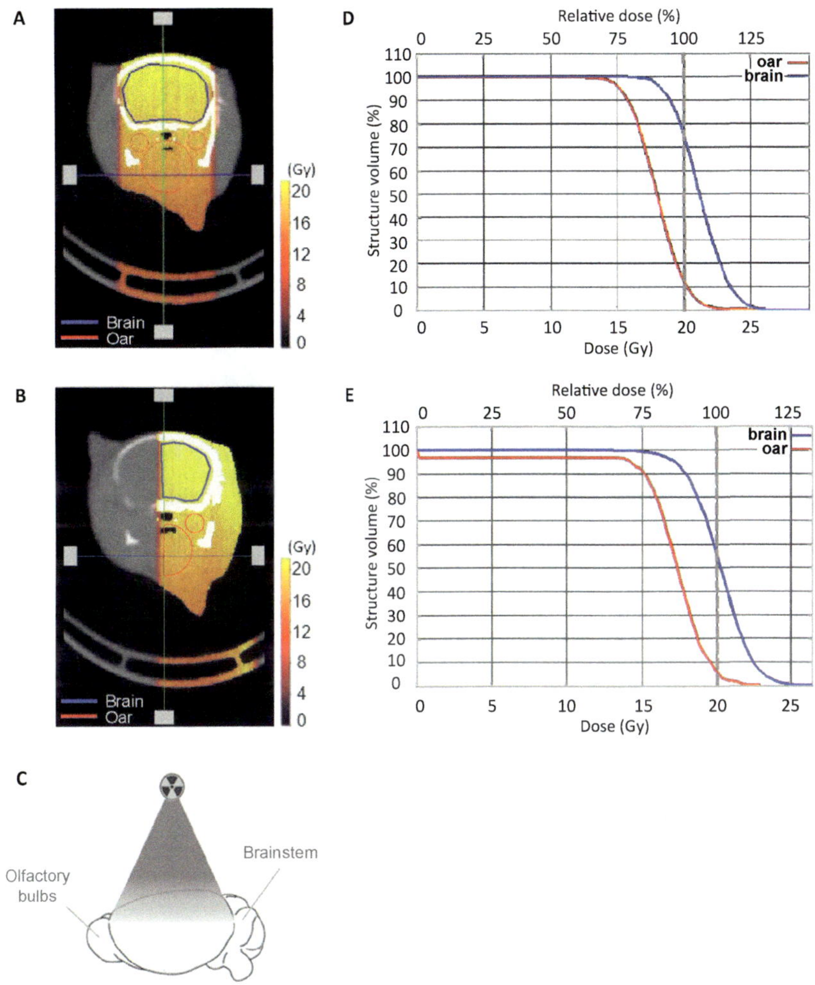

Fig. 1 X-ray beam parameters for brain irradiation of mouse using the SmARTPlan treatment planning software. (A) Dosimetry map representing x-ray whole-brain irradiation with a 20 Gy dose. The brain is outlined in blue and the organs at risk ((Oar) cervical lymph nodes, face muscles, salivary glands and large arteries) are outlined in red. (B) Dosimetry map representing x-ray hemispheric irradiation with a 20 Gy dose. The brain is outlined in blue and the organs at risk ((Oar) cervical lymph nodes, face muscles, salivary glands and large arteries) are outlined in red. (C) Schematic representation of how to set up the x-ray beam after the scanner control to avoid the olfactory bulbs and brainstem. (D) Dose-volume histogram illustrating the dose received for each structure of interest as a function of the dose set by the beam for the whole-brain irradiation. (E) Dose-volume histogram illustrating the dose received for each structure of interest as a function of the dose set by the beam for the hemispheric irradiation.

- The animal is manually restrained, held by the skin over the back of the neck and by the tail to prevent it from turning over. The animal is held upright to provide a good view of the cheek (Fig. 2A).
- A sterile single-use lancet is used with a 4 mm tip, specially made for "facial vein and submandibular blood samples" according to the supplier. It should be insert (and quickly withdraw) into the vessel located at the back of the cheek, just behind the point at which the upper and lower jawbones meet (Fig. 2B).

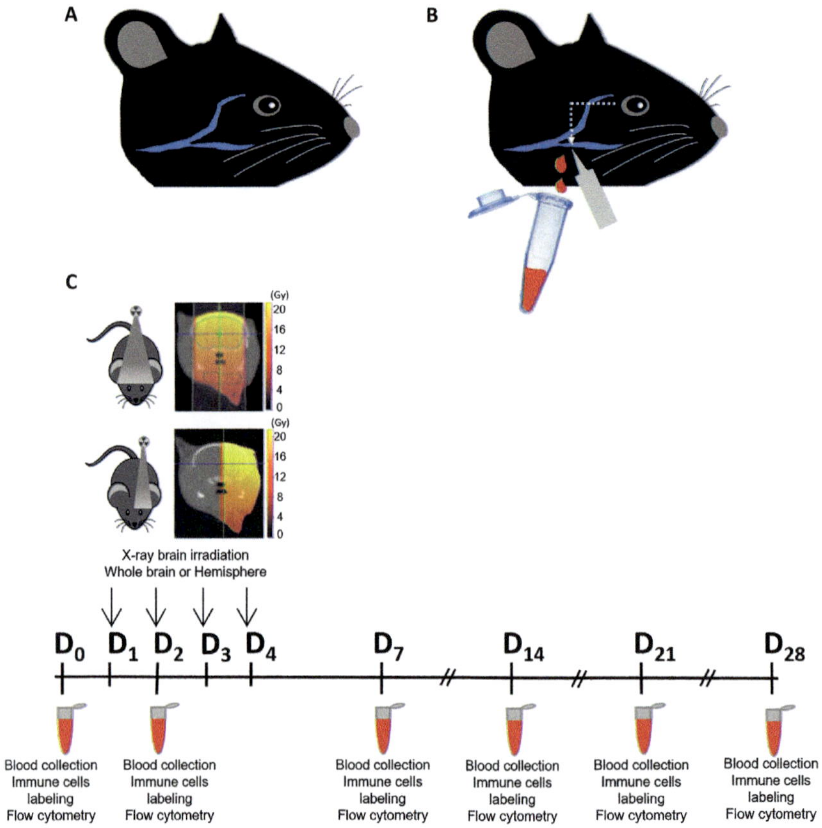

Fig. 2 Submandibular blood draw and repeatability of manipulation in a protocol for quantifying immune cell subtypes after brain x-ray irradiation. (A) Schematic representation of the facial and submandibular veins location. (B) Schematic representation of lancet positioning for blood sampling in the submandibular vein. (C) Experimental protocol designed for the investigation of the effects of x-ray by whole peripheral blood collection and circulating immune cells quantification before (D_0), during (D_2) and after (D_7, D_{14}, D_{21} and D_{28}) C57BL/6 mice brain irradiation. Blood was collected at six time points; samples were collected 2 days to 1 week apart.

- The blood is collected (drops are instantly visible) by quickly attaching a heparinized Eppendorf tube to the mouse cheek.

Note: after returning the animal to its cage, it will self-groom to clean any remnant blood from the fur. Observe the animal for any additional bleeding. If needed, serial blood samples can be collected by using the same sites, provided you allow sufficient time for recovery. Here, samples were taken at six different times before (D_0), during (D_2) and after (D_7, D_{14}, D_{21} and D_{28}) treatment (Fig. 2C).

3.4 FACS staining procedure

- Transfer 20 µL from blood sample onto a 96-wells V-bottom microplate. Group samples by panel (Table 2).
- Treat blood with 50 µL per well of 1× red blood cell lysis buffer.
- Centrifuge at room temperature (RT) at 300*g* for 5 min.

Note: be careful to balance the centrifuge properly, otherwise you risk losing sample volume and making the cell counting step biased.

- Repeat treating step with 50 µL per well of 1× red blood cell lysis buffer.
- Centrifuge at RT at 300*g* for 5 min.
- Wash with 100 µL per well of FACS buffer.
- Centrifuge at RT at 300*g* for 5 min.
- Treat with 50 µL per well of anti-CD16 and anti-CD32 antibodies (to prevent unspecific staining due to Fcγ receptors) (1/1000 in FACS buffer (Table 1)) for 15 min at RT.
- Centrifuge at RT at 300*g* for 5 min.
- Incubate cells with 50 µL per well of the extracellular mix containing the specific antibodies for 30 min at RT. The mix contains all extracellular antibodies at the correct dilution in the FACS buffer (Table 2).
- Centrifuge at RT at 300*g* for 5 min.
- Incubate cells with 50 µL per well with an aqua vivid live/dead cell marker (1/1000 in PBS (Table 1)) for 15 min at RT.
- Centrifuge at RT at 300*g* for 5 min.
- Fix cells with 80 µL per well with the fixation solution (3/4 diluent and 1/4 concentrate) and incubate for 30 min at RT.
- Centrifuge at RT at 300*g* for 5 min.
- Resuspend cells in 100 µL per well of FACS buffer.

Note: before transferring samples to the cytometer, plates can be stored at 4 °C, protected from light, for UP TO 5 days after the fixation step and the labeling can be further continued later. If cells are labeled with an intracellular antibody, they must be run to the cytometer within 4 h of the intracellular labeling step.

Table 2 Summary table of cytometry panel designed to identify each cell subtype of interest.

	Fluorochrome	Target	Dilution
Panel 1	FITC	NK1.1	1/200
	Percp/Cy5.5	B220	1/200
	PE	CD4	1/200
	PE/TR	RORgt	1/100
	PE/Cy7	CD8	1/200
	BV421	FoxP3	1/200
	BV510	Live/dead	1/1000
	BV605	CD3	1/100
	BV650	CD45.2	1/200
Panel 2	FITC	CD44	1/200
	Percp/Cy5.5	CD62L	1/200
	PE	PD-1	1/200
	PE/TR	CD4	1/200
	PE/Cy7	CD8	1/200
	BV421	CTLA4	1/100
	BV510	Live/dead	1/1000
	BV605	CD3	1/100
	BV650	CD45.2	1/200
Panel 3	FITC	F4/80	1/100
	Percp/Cy5.5	CD11b	1/500
	PE	CX3CR1	1/100
	PE/TR	SiglecF	1/200
	PE/Cy7	MHCII	1/500
	BV421	Ly6G	1/200
	BV510	Live/dead	1/1000
	BV605	Ly6C	1/200
	BV650	CD45.2	1/200

Panel 1 identified T cells as well as NK and B cells. Panel 2 identified T cell subtypes with exhaustion markers. Panel 3 identified myeloid cells and their subtypes.

- DO NOT centrifuge.
- Wash with 50 μL per well of 1× permeabilization buffer.
- Incubate with 50 μL per well of intracellular mix containing the specific antibodies for 1 h at RT. The mix contains all intracellular antibodies at the correct dilution in the permeabilization buffer (Table 2).

Note: be careful to the choice of antibodies when designing cytometry panels. The number of lasers and therefore colors available limit this choice (Table 3).

- DO NOT centrifuge.
- Wash with 50 μL per well of permeabilization buffer.
- Centrifuge at RT at 300*g* for 5 min.
- Resuspend cells in 100 μL per well of FACS buffer.
- Prior to the cytometer run, add 10 μL per well of counting beads.
- Run the samples to the cytometer.

Table 3 The CytoFLEX S features (Beckman Coulter).

	Lasers	Fluorescent channel	CytoFLEX channel names
WDM 1	*Purple* 405-nm	450/45 BP	PB450
		525/40 BP	KO252
		610/20 BP	Violet610
		660/20 BP	Violet660
WDM 2	*Yellow* 561-nm	610/20 BP	ECD mCherry
		585/42 BP	PE DsRed
		690/50 BP	PC5.5
		780/60 BP	PC7
WDM 3	*Red* 638-nm	660/20 BP	APC
		712/25 BP	APC-A700
		780/60 BP	APC-A750
	Blue 488-nm	525/40 BP	FITC GFP
		690/50 BP	PerCP PC5

The table summarizes the optical filter mount color codes. WDM = wavelength division multiplexer.

3.5 Data acquisition by flow cytometry and gating

Multispectral flow cytometry is performed on a CytExpert (Beckman Coulter) equipped with 405-, 488-, 561- and 638-nm lasers and 13 detectors for acquisition of fluorescence (4 detectors on the purple laser (450/45 BP, 525/40 BP, 610/20 BP, 660/20 BP), 4 on the yellow (610/20 BP, 585/42 BP, 690/50 BP, 780/60 BP), 3 on the red (660/20 BP, 712/25 BP, 780/60 BP) and 2 on the blue (525/40 BP, 690/50 BP)) (Table 3). The forward scatter (FSC) and side scatter (SSC) can also be detected. FSC provides information on cell size, while SSC at 90 °C provides information on cell granularity or structure. The CytExpert software (Beckman Coulter) controls instrument operation, collects and analyses data.

Note: depending on the cytometer filters, you cannot use the same fluorochromes.

- Build a compensation matrix for each panel: for each antibody used, prepare a 200 µL solution in a 5 mL polystyrene round-bottom tube with 150 µL of FACS buffer, 1 drop of compensation UltraComp eBeads™ and the fluorescent antibody at the correct dilution. The UltraComp eBeads™ contained unlabeled beads and beads labeled with the antibody, and are used to adjust lasers intensity and compensation.
- Before running sample to the cytometer, create a template file for each designed panel and save it on the cytometry software CytExpert. The template consists in performing the gating strategy you will use to identify all the cell subtypes (example in Fig. 3). The template file contains all dotplots displaying the targeted markers and therefore the populations to be identified.
- Import the compensation matrix.
- Set channels and labels of all antibodies used for each panel, determine the acquisition settings and the stopping rules (uncheck "events to record," set 600 s and 95 µL to acquire).
- Save the template using .XITM format.

Note: do not forget the compensation matrix step as it is essential for correcting the light intensity emitted by all the fluorochromes present in the panel as perceived by a photomultiplier tube (PMT).

- Create an Experiment and import the template for one panel.
- During acquisition preview, the cellular debris are excluded from the gate "Cells" by adjusting the FSC-A vs SSC-A plot, doublets are excluded from the gate "P1" by creating a combination of same-channel bivariate plots utilizing Area vs Height (FSC-A vs FSC-H). The gate "Single cells" is adjusted in the SSC-A vs SSC-H plot. Finally, alive CD45$^+$ are identified on CD45 vs Live/dead plots (Fig. 3A).

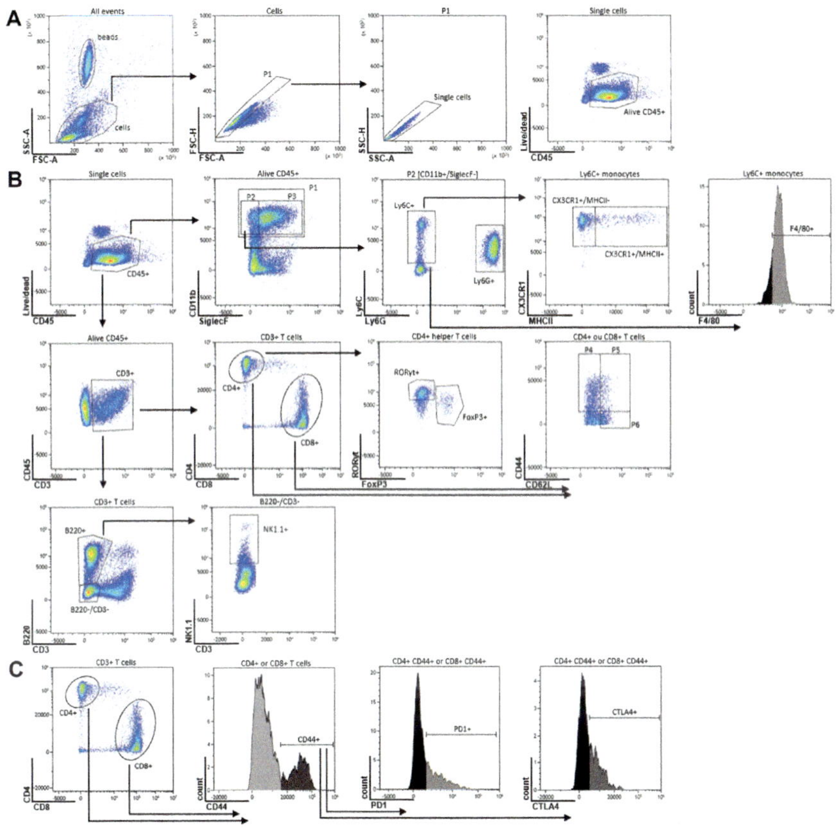

Fig. 3 Schematic gating strategy summarizing the analysis of preclinical cytometry data. (A) Gating strategy for alive $CD45^+$ identification after the exclusion of cellular debris, doublets and dead cells. (B) From alive hematopoietic cells ($dead^-CD45^+$), leukocyte's lineages were selected: myeloid ($CD11b^+$), T ($CD3^+$), B ($CD3^-B220^+$) and NK ($CD3^-NK1.1^+$) cells. Specific subtypes were then further identified: monocytes ($CD11b^+SiglecF^-Ly6C^+$), neutrophils ($CD11b^+SiglecF^-Ly6G^+$) and eosinophils ($CD11b^+SiglecF^+$) from myeloid cells; CD8 T cells ($CD45^+CD3^+CD8^+$), conventional CD4 T cells ($CD45^+CD3^+CD4^+$), regulatory T cells ($CD45^+CD3^+CD4^+FoxP3^+$) and T helper 17 cells ($CD45^+CD3^+CD4^+ROR\gamma t^+$) from T cells. (C) Effector ($CD4^+CD44^{high}CD62L^-$, $CD8^+CD44^{high}CD62L^-$), central memory ($CD4^+CD44^{high}CD62L^+$, $CD8^+CD44^{high}CD62L^+$) and naive ($CD4^+CD44^-CD62L^+$, $CD8^+CD44^-CD62L^+$) cells from $CD44^+CD4^+$ and $CD44^+CD8^+$ T cells as well as exhaustion markers (CTLA4 and PD-1) were determined.

- Adjust the flow rate <300 events/s for optimal resolution of fluorescence.
- Record and acquire 95 μL for each sample.
- Save the data using .FSC format.
- Proceed to data analysis using Kaluza® software (Beckman Coulter).

3.6 Data analysis and quantification

Quantification can be done in numbers of cells per microliter or as a percentage of a population according to the gating strategy (as an example, % $CD3^+$ among $CD45^+$ or % $CD4^+$ or $CD8^+$ among $CD3^+$). From whole peripheral blood, alive hematopoietic cells ($dead^-CD45^+$) were selected to gate on leukocyte's lineages: myeloid ($CD45^+CD11b^+$), T ($CD45^+CD3^+$), B ($CD45^+CD3^-B220^+$) and NK ($CD45^+CD3^-NK1.1^+$) cells allowing for the subpopulation dosage to be refined. From myeloid cells, monocytes ($CD11b^+SiglecF^-Ly6C^+$), neutrophils ($CD11b^+SiglecF^-Ly6G^+$) and eosinophils ($CD11b^+SiglecF^+$) were selected. From T lymphocytes, CD8 T cells ($CD3^+CD8^+$), conventional CD4 T cells ($CD3^+CD4^+FoxP3^-$), regulatory T cells ($CD3^+CD4^+FoxP3^+$) and T helper 17 cells ($CD3^+CD4^+ROR\gamma t^+$) were selected. From $CD4^+$ and $CD8^+$ T cells, effector ($CD4^+CD44^{high}CD62L^-$, $CD8^+CD44^{high}CD62L^-$), central memory ($CD4^+CD44^{high}CD62L^+$, $CD8^+CD44^{high}CD62L^+$) and naïve ($CD4^+CD44^-CD62L^+$, $CD8^+CD44^-CD62L^+$) cells were selected as well as exhaustion markers (CTLA4 and PD-1) (Fig. 3B and C). The number of cells per microliter can be calculated and normalized by the number of counting beads recorded.

Note: counting beads are supplied at an exact known concentration. As indicated above in the labeling protocol, 10 μL of counting beads are added in each well before the samples are run. As it is the same for each well, it can be named constant α. Thus, the number of beads experimentally recorded by the cytometer can be determined. The number of cells per microliter is obtained by the following formula:

$$¢/\mu L = \frac{\text{recorded } ¢^*\alpha}{\text{recorded beads}}/20$$

For each well, the calculation is the number of cells recorded by the cytometer (first plot Fig. 3A) multiplied by the constant α and divided by the number of beads recorded by the cytometer. This value is divided by the 20 μL of blood deposited in each sample.

4. Concluding remarks/notes

Using flow cytometry, this method brings an absolute, reproducible and reliable method to follow with time the variation of circulating myeloid and lymphoid subtypes after brain exposure to irradiation. It is a single-cell analysis method reflecting the systemic inflammation. This very sensitive method permits the quantification of several markers simultaneously at the single cell level. In the brain cancer field, it is essential to identify and understand the evolution of leukocyte subsets after brain exposure given the implication of inflammation in the tumor response to radiotherapy.

Acknowledgments

Region Normandie (to J.C.). Ligue contre le Cancer (to S.V.). CNRS (Centre National de la Recherche Scientifique) through the 80|Program (to S.V.). Animal care facility (CURB@CYCERON, Caen, France). Flow cytometry facility (Normandie Equine Vallee platform, LABEO BIOTARGEN, Saint-Contest, France).

Conflict of interest statement

The authors declare no conflict of interest.

References

Ahn, S., et al. (2020). The association between total lymphocyte count after concomitant chemoradiation and overall survival in patients with newly diagnosed glioblastoma. *Journal of Clinical Neuroscience*, *71*, 21–25.

Akkari, L., et al. (2020). Dynamic changes in glioma macrophage populations after radiotherapy reveal CSF-1R inhibition as a strategy to overcome resistance. *Science Translational Medicine*, *12*, eaaw7843.

Alifieris, C., & Trafalis, D. T. (2015). Glioblastoma multiforme: Pathogenesis and treatment. *Pharmacology & Therapeutics*, *152*, 63–82.

Campian, J. L., Sarai, G., Ye, X., Marur, S., & Grossman, S. A. (2014). Association between severe treatment-related lymphopenia and progression-free survival in patients with newly diagnosed squamous cell head and neck cancer. *Head & Neck*, *36*, 1747–1753.

Cesaire, M., Le Mauff, B., Rambeau, A., Toutirais, O., & Thariat, J. (2020). Mechanisms of radiation-induced lymphopenia and therapeutic impact. *Bulletin du Cancer*, *107*, 813–822.

Chongsathidkiet, P., et al. (2018). Sequestration of T cells in bone marrow in the setting of glioblastoma and other intracranial tumors. *Nature Medicine*, *24*, 1459–1468.

Collet, S., et al. (2021). Simultaneous mapping of vasculature, hypoxia, and proliferation using dynamic susceptibility contrast MRI, 18F-FMISO PET, and 18F-FLT PET in relation to contrast enhancement in newly diagnosed glioblastoma. *Journal of Nuclear Medicine*, *62*, 1349–1356.

Dutoit, V., et al. (2020). Impact of radiochemotherapy on immune cell subtypes in high-grade glioma patients. *Frontiers in Oncology*, *10*, 89.

Francisco, C. C., Howarth, G. S., & Whittaker, A. L. (2015). Effects on animal wellbeing and sample quality of 2 techniques for collecting blood from the facial vein of mice. *Journal of the American Association for Laboratory Animal Science*, *54*, 76–80.

Ghosh, S., et al. (2023). Radiation-induced circulating myeloid-derived suppressor cells induce systemic lymphopenia after chemoradiotherapy in patients with glioblastoma. *Science Translational Medicine*, *15*, eabn6758.

Golde, W. T., Gollobin, P., & Rodriguez, L. L. (2005). A rapid, simple, and humane method for submandibular bleeding of mice using a lancet. *Lab Animal*, *34*, 39–43.

Grossman, S. A., et al. (2011). Immunosuppression in patients with high-grade gliomas treated with radiation and temozolomide. *Clinical Cancer Research*, *17*, 5473–5480.

Grossman, S. A., et al. (2015). Survival in patients with severe lymphopenia following treatment with radiation and chemotherapy for newly diagnosed solid tumors. *Journal of the National Comprehensive Cancer Network*, *13*, 1225–1231.

Hammi, A., Paganetti, H., & Grassberger, C. (2020). 4D blood flow model for dose calculation to circulating blood and lymphocytes. *Physics in Medicine and Biology, 65*, 055008.

Han, S., et al. (2014). Tumour-infiltrating CD4(+) and CD8(+) lymphocytes as predictors of clinical outcome in glioma. *British Journal of Cancer, 110*, 2560–2568.

Huang, J., et al. (2015). Clinical and dosimetric predictors of acute severe lymphopenia during radiation therapy and concurrent temozolomide for high-grade glioma. *International Journal of Radiation Oncology, Biology, Physics, 92*, 1000–1007.

Jarosz-Biej, M., Smolarczyk, R., Cichoń, T., & Kułach, N. (2019). Tumor microenvironment as a 'game changer' in cancer radiotherapy. *International Journal of Molecular Sciences, 20*.

Jo, E. J., Bae, E., Yoon, J.-H., Kim, J. Y., & Han, J. S. (2021). Comparison of murine retroorbital plexus and facial vein blood collection to mitigate animal ethics issues. *Laboratory Animal Research, 37*, 12.

Klemm, F., et al. (2020). Interrogation of the microenvironmental landscape in brain tumors reveals disease-specific alterations of immune cells. *Cell, 181*, 1643–1660.e17.

Kumar, N., et al. (2020). Impact of volume of irradiation on survival and quality of life in glioblastoma: A prospective, phase 2, randomized comparison of RTOG and MDACC protocols. *Neurooncology Practice*, 7, 86–93.

Leblond, M. M., et al. (2016). Hypoxia induces macrophage polarization and re-education toward an M2 phenotype in U87 and U251 glioblastoma models. *Oncoimmunology, 5*, e1056442.

Leblond, M. M., et al. (2017). M2 macrophages are more resistant than M1 macrophages following radiation therapy in the context of glioblastoma. *Oncotarget, 8*, 72597–72612.

Macintyre, A. N., et al. (2021). Long-term recovery of the adaptive immune system in rhesus macaques after total body irradiation. *Advances in Radiation Oncology, 6*, 100677.

Mohan, R., et al. (2020). Proton therapy reduces the likelihood of high-grade radiation-induced lymphopenia in glioblastoma patients: Phase II randomized study of protons vs. photons. *Neuro-Oncology, 23*, 284–294.

Nakamura, N., Kusunoki, Y., & Akiyama, M. (1990). Radiosensitivity of CD4 or CD8 positive human T-lymphocytes by an in vitro colony formation assay. *Radiation Research, 123*, 224–227.

Pham, T.-N., et al. (2023). Beyond lymphopenia, unraveling radiation-induced leucocyte subpopulation kinetics and mechanisms through modeling approaches. *Journal of Experimental & Clinical Cancer Research, 42*, 50.

Pham, T.-N., et al. (2024). Early effects of different brain radiotherapy modalities on circulating leucocyte subpopulations in rodents. *International Journal of Radiation Biology, 1–12.*

Prosniak, M., et al. (2013). Glioma grade is associated with the accumulation and activity of cells bearing M2 monocyte markers. *Clinical Cancer Research, 19*, 3776–3786.

Quail, D. F., & Joyce, J. A. (2017). The microenvironmental landscape of brain tumors. *Cancer Cell, 31*, 326–341.

Rafat, M., et al. (2018). Macrophages promote circulating tumor cell-mediated local recurrence following radiotherapy in immunosuppressed patients. *Cancer Research, 78*, 4241–4252.

Tomaszewski, W., Sanchez-Perez, L., Gajewski, T. F., & Sampson, J. H. (2019). Brain tumor microenvironment and host state: Implications for immunotherapy. *Clinical Cancer Research, 25*, 4202–4210.

Wang, S.-J., & Haffty, B. (2018). Radiotherapy as a new player in immuno-oncology. *Cancers (Basel), 10*, 515.

Yovino, S., Kleinberg, L., Grossman, S. A., Narayanan, M., & Ford, E. (2013). The etiology of treatment-related lymphopenia in patients with malignant gliomas: Modeling radiation dose to circulating lymphocytes explains clinical observations and suggests methods of modifying the impact of radiation on immune cells. *Cancer Investigation, 31*, 140–144.

CHAPTER TEN

Flow cytometry-assisted analysis of phenotypic maturation markers on an immortalized dendritic cell line

Ginevra Campia[a], Manuel Beltrán-Visiedo[a], Ruth Soler-Agesta[a,b], Ai Sato[a], Norma Bloy[a], Liwei Zhao[c,d], Peng Liu[c,d], Oliver Kepp[c,d], Guido Kroemer[c,d,e], Lorenzo Galluzzi[a,f,g,*], and Claudia Galassi[a,*]

[a]Department of Radiation Oncology, Weill Cornell Medical College, New York, NY, United States
[b]University of Zaragoza/Aragón Health Research Institute, Biochemistry and Molecular and Cell Biology, Zaragoza, Spain
[c]Equipe Labellisée Par La Ligue Contre Le Cancer, Université de Paris, Sorbonne Université, Centre de Recherche des Cordeliers, Institut Universitaire de France, Paris, France
[d]Metabolomics and Cell Biology Platforms, Gustave Roussy, Université Paris Saclay, Villejuif, France
[e]Pôle de Biologie, Hôpital européen Georges Pompidou, AP-HP, Paris, France
[f]Sandra and Edward Meyer Cancer Center, New York, NY, United States
[g]Caryl and Israel Englander Institute for Precision Medicine, New York, NY, United States
*Corresponding authors: e-mail address: deadoc80@gmail.com; clg4005@med.cornell.edu

Contents

Methods in Cell Biology, Volume 189
ISSN 0091-679X
https://doi.org/10.1016/bs.mcb.2024.05.008

Abstract

Dendritic cells (DCs), and especially so conventional type I DCs (cDC1s), are fundamental regulators of anticancer immunity, largely reflecting their superior ability to engulf tumor-derived material and process it for cross-presentation on MHC Class I molecules to $CD8^+$ cytotoxic T lymphocytes (CTLs). Thus, investigating key DC functions including (but not limited to) phagocytic capacity, expression of CTL-activating ligands on the cell surface, and cross-presentation efficacy is an important component of multiple immuno-oncology studies. Unfortunately, DCs are terminally differentiated cells, implying that they cannot be propagated indefinitely *in vitro* and hence must be generated *ad hoc* from circulating or bone marrow-derived precursors, which presents several limitations. Here, we propose a simple, cytofluorometric method to quantify phenotypic activation markers including CD80, CD86 and MHC class II molecules on the surface of a conditionally immortalized immature DC line that can be indefinitely propagated *in vitro* but also driven into maturation at will with a simple change in culture conditions. Upon appropriate scaling and automatization, this approach is compatible with high-throughput screening programs for the discovery of novel DC activators that do not suffer from batch variability and other limitations associated with the generation of fresh DCs.

1. Introduction

Dendritic cells (DCs) are major regulators of the interaction between the innate and adaptive branch of the immune response (Cabeza-Cabrerizo et al., 2021). Indeed, immature DCs constantly probe their surroundings by constantly taking up soluble material and (relatively small) particulate, processing it in the endosomal compartment and presenting it on the cell surface in association with MHC molecules (Blander, 2018; Jhunjhunwala et al., 2021; Yang et al., 2023). DCs—especially so-called type I conventional DCs (cDC1s) (Hildner et al., 2008; Johnson et al., 2022; Theisen et al., 2018)—are particularly efficient antigen-presenting cells (APCs) as they display peptides derived from extracellular material not only on MHC Class II molecules for presentation to $CD4^+$ helper (T_H) cells (as most APCs do) but also on MHC Class I molecules for presentation to $CD8^+$ cytotoxic T lymphocytes (CTLs), a process that is commonly known as antigen cross-presentation (Blander, 2018; Cruz et al., 2017; Yin et al., 2021). Cross-presentation (which generally occurs in secondary or tertiary lymphoid organs) is particularly important for the activation of both natural and therapy-driven tumor-targeting immune responses against malignant cells (Heras-Murillo et al., 2024; Jhunjhunwala et al., 2021; Woo et al., 2015; Yatim et al., 2017). In line with this notion, considerable preclinical and clinical efforts are being dedicated at the development of immunotherapeutic strategies that, directly or indirectly, harness the immunostimulatory potential of DCs (Kyrysyuk & Wucherpfennig, 2023; Palucka & Banchereau, 2012), including (but not limited to) approaches involving the administration of exogenous or release of endogenous DC stimulators such as Toll-like receptor (TLR) agonists (Hennessy et al., 2010; Le Naour & Kroemer, 2023; Vacchelli et al., 2013a; Vanpouille-Box et al., 2019) and immunostimulatory cytokines (Deckers et al., 2023; García-Martínez et al., 2018; Mempel et al., 2024), as well as the use of patient-derived DCs as therapeutic vaccines upon *ex vivo* expansion and exposure to autologous or heterologous tumor-derived material plus immunostimulatory cues (Cultrara et al., 2023; Fucikova et al., 2022; Harari et al., 2020; Vacchelli et al., 2013b; Wculek et al., 2020).

To elicit potent anticancer immune responses, DCs including cDC1s must expose not only MHC Class I and II molecules, but also a series of T cell costimulating ligands on their surface, including (but not limited to) CD80, CD83 and CD86 (Jhunjhunwala et al., 2021). Such a mature configuration—which is generally associated with the expression of chemokine receptors guiding DCs to secondary or tertiary lymphoid organs (Worbs et al., 2017)—is indeed fundamental for (cross)presentation to result in robust T cell priming rather than anergy, which would instead promote immunological tolerance (Jhunjhunwala et al., 2021; Kroemer et al., 2022). Importantly, DC maturation can be elicited by a number of signals that operate as pattern recognition receptor (PRR) agonists, including pathogen-derived molecules commonly known as microbe-associated molecular patterns (MAMPs), as well as endogenous molecules that are released or exposed by stressed and dying (cancer) cells, which are commonly referred to as damage-associated molecular patterns (DAMPs) (Brubaker et al., 2015; Chen & Nuñez, 2010; Gong et al., 2020; Kroemer et al., 2022; Marchi et al., 2023). The spatiotemporally ordered emission of DAMPs by dying cancer cells is fundamental for their death to be perceived by DCs as immunogenic, and hence culminate with the initiation of a tumor-specific immune response (Kroemer et al., 2022).

Based on these observations, it is not surprising that numerous immuno-oncology studies involve the assessment of one or more DC functions. Unfortunately, DCs are terminally differentiated cells, implying that they cannot be propagated indefinitely *in vitro* but must be purified or generated *ad hoc* from circulating or bone marrow-derived precursors (Cerrato et al., 2020; Hong & Dobrovolskaia, 2024; Isnard et al., 2020). Besides being laborious and associated with a relatively poor yield, this approach suffers from considerable batch variability, limiting its applicability to high-throughput screening platforms. A conditionally immortalized mouse DC line that can be propagated indefinitely *in vitro* but also differentiated at will *via* minimal changes in culture conditions that enable the expression of a potent oncogene (SV40 large T antigen) coupled to the repression of tumor protein p53 (TRP53, best known as p53) can at least in part circumvents this problem, *de facto* offering a tool that can be scalable for the high-throughput screening of novel molecules that influence DC functions (Zhao et al., 2021).

Here, we propose a simple, flow cytometry-assisted approach to quantify phenotypic activation markers including CD80, CD86 and MHC class II molecules on these conditionally immortalized DCs optionally exposed to maturation conditions.

2. Materials

2.1 Common disposables (*see Note 1*)

- Corning™ 96-Well, Cell Culture-Treated, V-Shaped-Bottom Microplate (#3897, Corning)
- Falcon® 12-well Clear Flat Bottom TC-treated Multiwell Cell Culture Plate, with Lid, Individually Wrapped, Sterile (#353043, Corning)
- Falcon® 15 mL High Clarity PP Centrifuge Tube, Conical Bottom, with Dome Seal Screw Cap, Sterile (#352196, Corning)
- Falcon® 5 mL Round-Bottom Polypropylene Test Tubes Without Cap (#352002, Corning)
- Falcon® 50 mL High Clarity PP Centrifuge Tube, Conical Bottom, Sterile (#352070, Corning)
- Nunc™ EasYFlask™ 75 cm^2 Cell Culture Flasks (#156499, Thermo Scientific)

2.2 Cells and reagents (*see Note 1*)

- APC/Cyanine7 anti-mouse CD11c Antibody (clone N418) (#117324, BioLegend)
- Bovine Serum Albumin (BSA) Powder—Low-Electrolyte Grade (#700-100P, GeminiBio)
- Corning™ 500 mL RPMI 1640 1× (#15-040-CV, Corning)
- Corning™ Phosphate-Buffered Saline (PBS), 1× without calcium and magnesium, pH 7.4 ± 0.1 (#21-040-CMR, Corning)

- Dexamethasone (Dex) (#02190040-*CF*, MP Biomedicals) (*see Note 2*)
- Doxycycline hyclate (Dox) (#D9891, Sigma-Aldrich) (*see Note 2*)
- eBioscience™ Foxp3/Transcription Factor Fixation/Permeabilization Concentrate and Diluent (#00-5521-00, Thermo Scientific)
- eBioscience™ Permeabilization Buffer (10×) (#00-8333-56, Thermo Scientific)
- FITC anti-mouse CD86 Antibody (clone GL-1) (#105006, BioLegend)
- Foundation™ Fetal Bovine Serum (FBS) (#900-108-500, GeminiBio) (*see Note 3*)
- Gibco™ 2-Mercaptoethanol (#21985023, Thermo Scientific) (*see Note 4*)
- Gibco™ Gentamicin (50 mg/mL) (#15750078, Thermo Scientific)
- Gibco™ HEPES (1 M) (#15630080, Thermo Scientific)
- Gibco™ Penicillin-Streptomycin-Glutamine (100×) (#10378016, Thermo Scientific)
- Gibco™ Sodium Pyruvate (100 mM) (#11360070, Thermo Scientific)
- Gibco™ Trypsin-EDTA (0.5%), no phenol red (#15400054, Thermo Scientific)
- Inducible immortalized DCs (iniDCs) (Richter et al., 2013)
- PE/Cyanine7 anti-mouse CD80 Antibody (clone 16-10A1) (#104734, BioLegend)
- PerCP/Cyanine5.5 anti-mouse I-A/I-E Antibody (clone M5/114.15.2) (#107625, BioLegend)
- Recombinant Murine GM-CSF (#315-03, PeproTech) (*see Note 5*)
- TLR4 Agonist—Ultrapure lipopolysaccharide (LPS) from *E. coli* 055:B5 (#tlrl-pb5lps, InvivoGen)
- TruStain FcX™ (anti-mouse CD16/32) Antibody (#101319, BioLegend)
- Zombie aqua™ Fixable Viability Kit (#423102, BioLegend)

2.3 Equipment (*see Note 1*)

- Benchtop centrifuge, such as Allegra X-14R Centrifuge, 120 V, 60 Hz (#A99464, Beckman Coulter)
- Flow cytometer equipped with lasers for acquisition of SSC-H, FSC-H, BL1-H (488 nm laser, 530/30 filter), BL3-H (488 nm laser, 695/40 filter), BL4-H (488 nm laser, 780/60 filter), RL3-H (637 nm laser, 780/60 filter), VL2-H (405 nm laser, 512/25 filter), such as Attune NxT flow cytometer (Thermo Scientific).
- Humidified cell culture incubator, such as Heracell™ VIOS 160i CO2 Incubator (#51030285, Thermo Scientific)

- Laboratory biosafety cabinet (Class II), such as Purifier Logic+ Class II Type A2 Biological Safety Cabinet (#320419101, Labconco)

2.4 Software

- FlowJo v. 10.9.0 (FlowJo LLC)
- Excel (Microsoft)
- Prism v 10.0.1 (GraphPad)

3. Methods

3.1 *In vitro* de-immortalization and maturation of iniDCs

1. iniDCs are routinely expanded at 37 °C under 5% CO_2 in Nunc™ EasYFlask™ 75 cm^2 Cell Culture Flasks containing 22 mL RPMI 1640 supplemented with 10% FBS, 100 U/mL penicillin, 100 µg/mL streptomycin, 0.3 mg/mL glutamine, 50 µg/mL gentamicin, 5 mM HEPES, 1 mM sodium pyruvate and 50 µM 2-mercaptoethanol (complete culture medium), further supplemented with 50 µM 2-mercaptoethanol, 10 ng/mL recombinant GM-CSF, 100 nM Dex and 1 µg/mL Dox (to preserve immortalization).
2. Once iniDCs reach 70–80% confluence, exhausted culture medium is transferred to a 50 mL Falcon® tube and cells are gently washed with 5 mL pre-warmed PBS at 37 °C.
3. PBS is discarded and iniDCs are detached by adding 10 mL Gibco™ Trypsin-EDTA (0.5%) solution for 5 min at 37 °C under 5% CO_2 (*see Note 6*).
4. Once iniDCs are detached (*see Note 7*), Gibco™ Trypsin-EDTA (0.5%) solution is neutralized with 5 mL exhausted culture media and the cell suspension is transferred to a 50 mL Falcon® tube.
5. IniDCs are pelleted by centrifugation for 5 min at 300*g*, followed by aspiration of supernatants and resuspension in 1 mL complete culture medium.
6. To propagated immortalized iniDCs, 5×10^6 cells are seeded in a new Nunc™ EasYFlask™ 75 cm^2 Cell Culture Flask with 22 mL pre-warmed complete culture medium supplemented with 50 µM 2-mercaptoethanol, 10 ng/mL recombinant GM-CSF, 100 nM Dex and 1 µg/mL Dox.
7. To generate de-immortalized iniDCs (de-iniDCs), 7×10^6 cells are seeded in a new Nunc™ EasYFlask™ 75 cm^2 Cell Culture Flask with 22 mL pre-warmed complete culture medium supplemented with

50 μM 2-mercaptoethanol and 20 ng/mL recombinant GM-CSF. Complete de-immortalization takes 3–5 days (*see Notes 8 and 9*).

8. For flow cytometry experiments, iniDCs or de-iniDCs are detached as detailed in steps 2–5, resuspended in complete medium (for iniDCs supplemented with 50 μM 2-mercaptoethanol, 10 ng/mL recombinant GM-CSF, 100 nM Dex and 1 μg/mL Dox; for de-iniDCs supplemented with 50 μM 2-mercaptoethanol and 20 ng/mL recombinant GM-CSF) at a density of 3×10^5 cells/mL and seeded in Falcon® 12-well Clear Flat Bottom TC-treated Multiwell Cell Culture Plates (1 mL/well)
9. Maturation of iniDCs and de-iniDCs is induced by adding 100 ng/mL ultrapure LPS to the culture medium for 24–72 h or any potential compound of interest (*see Note 10*).

3.2 Staining

1. Exhausted culture medium is transferred to a Falcon® 5 mL Round-Bottom Polypropylene Test Tube.
2. PBS is discarded and cells are detached by adding 1 mL Gibco™ Trypsin-EDTA (0.5%) solution for 5 min at 37 °C under 5% CO_2 (*see Note 11*).
3. Once cells are detached (*see Note 7*) Gibco™ Trypsin-EDTA (0.5%) solution is neutralized with 1 mL exhausted culture media and the cell suspension is transferred to a Falcon® 5 mL Round-Bottom Polypropylene Test Tube.
4. Cells are pelleted by centrifugation for 5 min at 300 g, followed by aspiration of supernatants and resuspension in 200 μL pre-warmed PBS and transferred to Corning™ 96-Well, Cell Culture-Treated, V-Shaped-Bottom Microplate (*see Note 12*).
5. Plates are centrifuged for 5 min at 300 *g* at room temperature and supernatants are discarded by inversion.
6. Each well is dispensed with 25 μL Zombie aqua™ Fixable Viability Kit (1:200 in PBS) and plates are incubated at 4 °C in the dark for 15 min, followed by rinsing with 175 μL pre-warmed PBS.
7. Plates are centrifuged for 5 min at 300 *g* at room temperature and supernatants are discarded by inversion.
8. Each well is dispensed with 25 μL Murine TruStain FcX™ (1:200 in PBS supplemented with 0.5% (*w*/*v*) BSA) and plates are incubated at 4 °C in the dark for 15 min.
9. Each well is dispensed with 25 μL PBS supplemented with 0.5% (w/v) BSA plus 1:25 (v:v) PE/Cyanine7 anti-mouse CD80 Antibody, 1:25

(v:v) FITC anti-mouse CD86 Antibody, 1:25 (v:v) APC/Cyanine7 anti-mouse CD11c Antibody and 1:312 (v:v) PerCP/Cyanine5.5 anti-mouse I-A/I-E Antibody and plates are incubated at 4 °C in the dark for 25 min (*see Notes 13 and 14*).

10. Each well is washed with 150 μL PBS supplemented with 0.5% (w/v) BSA and plates are centrifuged for 5 min at 300 *g* at room temperature.
11. Supernatants are discarded by inversion and cells are resuspended in 200 μL eBioscience™ Foxp3/Transcription Factor Fixation/Permeabilization Concentrate and Diluent, followed by plate incubation at 4 °C in the dark for 30 min (*see Note 15*).
12. Plates are centrifuged for 5 min at 300 *g* at 4 °C, supernatants are discarded by inversion, and cells are resuspended in 200 μL 1× eBioscience™ Permeabilization Buffer.
13. Finally, plates are centrifuged for 5 min at 300 *g* at 4 °C, supernatants are discarded by inversion, and cells are resuspended in 200 μL PBS supplemented with 0.5% (w/v) BSA.

3.3 Acquisition and gating

1. Laser voltage (gain) and compensation are set on unstained cells as well as cells stained with each fluorochrome separately (*see Note 13*).
2. Acquire >10,000 events per sample, ideally at a flow rate <500 events/s (*see Notes 16 and 17*).
3. Gating should be performed as followed (Fig. 1):
   a. Cells are identified on FSC-H vs SSC-H dot plots, excluding events with low FSC-H and SSC-H (debris) (Fig. 1A).
   b. On cells, singlets are identified on FSC-A vs FSC-H dot plots, including events exhibiting a direct proportionality between these two parameters (Fig. 1B).
   c. On singlets, live cells are identified on FSC-H vs VL2-H dots plots, including events that stain negatively for Zombie aqua™ (Fig. 1C).
   d. On live cells, DCs are identified on FSC-H vs RL3-H dot plots, including events that stain positively for APC/Cy7 (CD11c$^+$) based on fluorescence minus one (FMO) controls (*see Note 13*) (Fig. 1D).
   e. On CD11c$^+$ DCs, MHC Class II-expressing cells are identified on FSC-H vs BL3-H dot plots, including events that stain positively for PerCP/Cy7 (MHC II$^+$) based on FMO controls (*see Note 13*) (Fig. 1E).

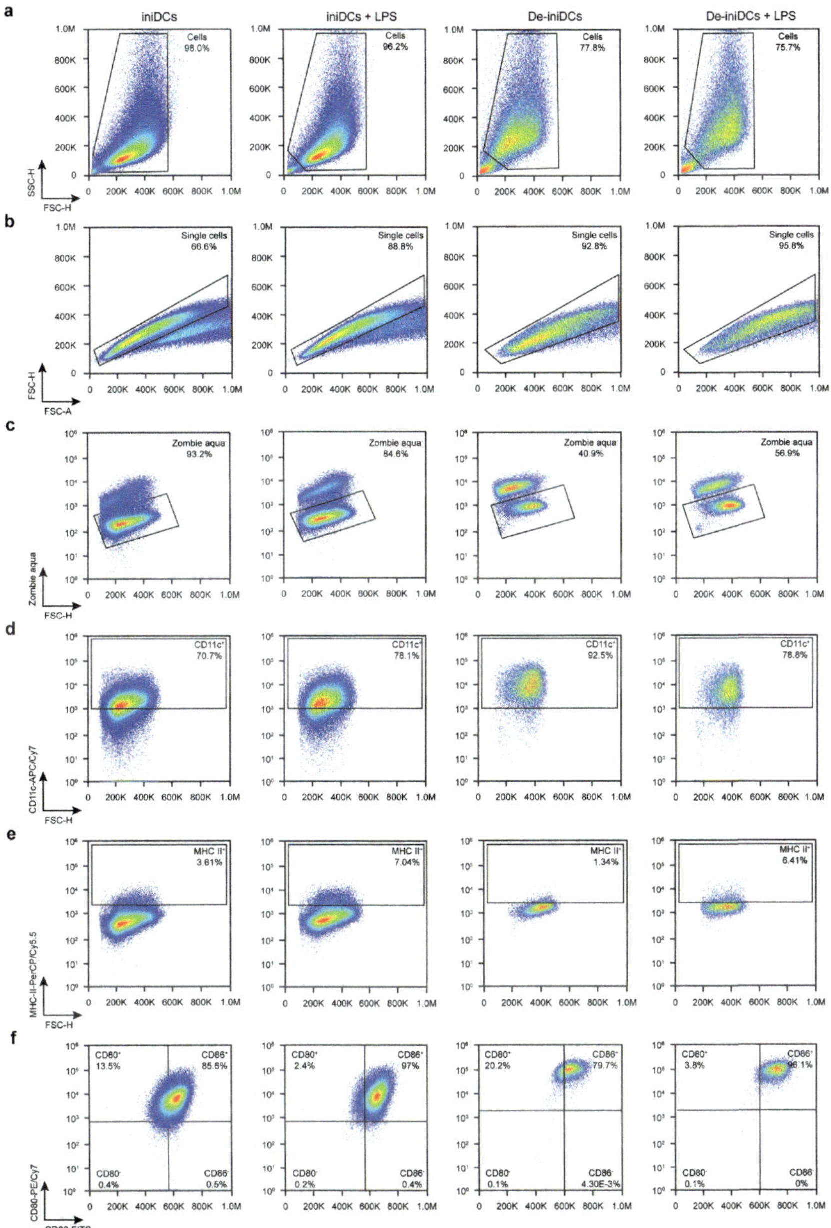

Fig. 1 Gating strategy. iniDCs were maintained in culture, optionally converted to de-iniDCs and exposed to lipopolysaccharide as detailed in Section 3.1, then stained with antibodies specific for CD11c, CD80, CD86 and MHC Class II molecules plus Zombie aqua as detailed in Section 3.2, followed by fluorescence acquisition and gating as detailed in Section 3.3. (A) Debris exclusion. (B) Singlet gating on cells. (C) Zombie aqua^{-} (live) cell gating on singlets. (D) CD11c^{+} (DC) cell gating on live cells. (E) MHC II^{+} cell identification on DCs. (F) CD80^{+} and/or CD86^{+} cell identification on DCs. Representative dot plots are shown, numbers indicate the percentage of events within each gate.

- **f.** On CD11c^{+} DCs, CD80- and/or CD86-expressing cells are identified on BL1-H vs BL4-H dot plots, including events that stain positively for FITC (CD86^{+}) and/or Pe/Cy7 (CD80^{+}) based on FMO controls (*see Note 13*) (Fig. 1F)
- **g.** Data can be represented as percentage of positive cells (Figs. 1 and 2) and/or mean fluorescence intensity (MFI) for the entire cell population of interest (Fig. 2).

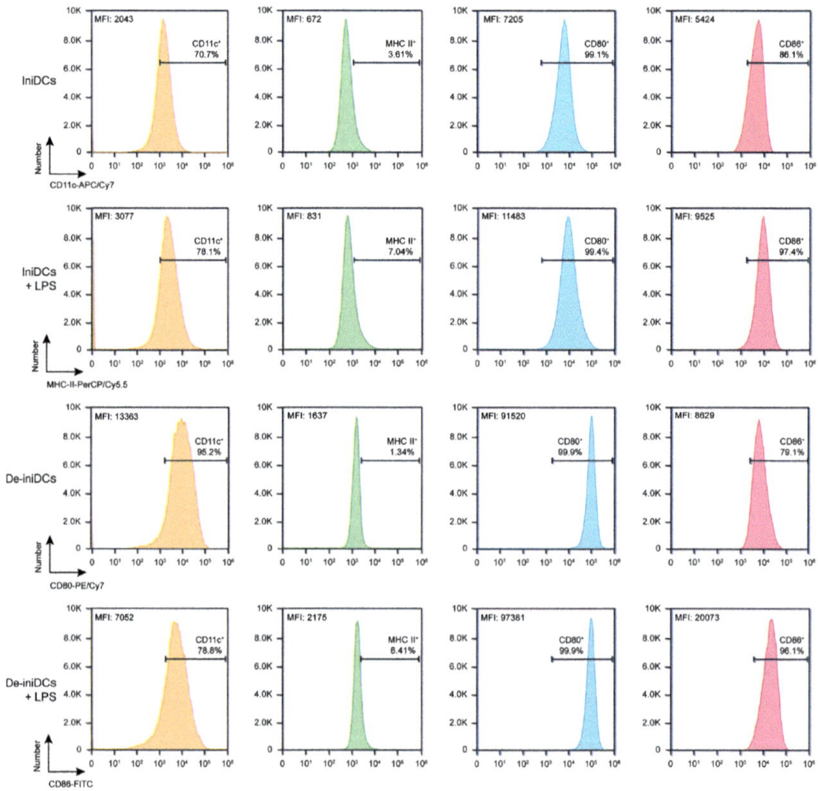

Fig. 2 Phenotypic maturation markers on iniDCs and de-iniDCs. iniDCs were maintained in culture, optionally converted to de-iniDCs and exposed to lipopolysaccharide as detailed in Section 3.1, then stained with antibodies specific for CD11c, CD80, CD86 and MHC Class II molecules plus Zombie aqua as detailed in Section 3.2, followed by fluorescence acquisition and gating as detailed in Section 3.3. Representative histograms are shown, numbers indicate the percentage of events within each gate and MFI for the entire cell population.

4. Notes

1. Virtually equivalent disposables, reagents and equipment can be purchased from alternative sources, generally at comparable cost.
2. The Dex and Dox stock solution (100 μM Dex plus 1 mg/mL Dox in PBS) is stable at 4 °C for a maximum of 3 months.
3. Prior to first use, FBS must be decomplemented at 56 °C for 30 min in a water bath. This prevents the potential activation of the complement system (Mastellos et al., 2024), which may have toxic effects of iniDCs and de-iniDCs.
4. 2-Mercaptoethanol can cause respiratory tract and skin irritations, and hence it should be manipulated with appropriate certified personal protective equipment (PPE).
5. Recombinant GM-CSF is stable at −20 °C but repeated freeze-thawing cycles should be avoided to minimize degradation.
6. Gibco™ CTS™ TrypLE™ Select Enzyme (#A1285901, Thermo Scientific) or other commercially available reagents can be employed as alternatives to Gibco™ Trypsin-EDTA (0.5%) solution.
7. As FBS leftover may inhibit the enzymatic activity of the Gibco™ Trypsin-EDTA (0.5%) solution, it is recommendable to visually ensure complete detachment by light microscopy before proceeding.
8. De-iniDCs stop proliferating 48 h upon Dex/Dox withdrawal.
9. De-iniDCs are larger than iniDCs and tend to exhibit increased adherence to the culture substrate, a phenotype that is partially lost upon LPS stimulation (*see also Note 11*)
10. IniDC maturation can be achieved as early as 12 h after incubation with the appropriate dose of LPS.
11. Falcon® Cell scrapers (#353085, Corning) may also be used to detach de-iniDCs after incubation in Gibco™ Trypsin-EDTA (0.5%) solution (*see also Note 9*).
12. Falcon® 5 mL Round-Bottom Polypropylene Test Tubes can also be employed for staining, adjusting total reagent volume to 200–250 μL per sample.
13. It is fundamental to include samples stained with single fluorochromes as well as FMO samples to adjust gain, compensation, and gating.
14. Staining samples in the dark is important to minimize fluorochrome photobleaching, which may negatively affect acquisition.

15. This step can also be performed at 4 °C in the dark overnight.
16. The acquisition of a high number of events per sample is important to enable proper gating and analysis.
17. An excessive flow rate may result in suboptimal acquisition, in turn negatively affecting gating and analysis.

5. Concluding remarks

The protocol described herein provides a straightforward approach to harness iniDCs and their de-immortalized counterparts to cytofluorometrically quantify phenotypic DC maturation markers including CD80, CD86 and MHC Class II molecules. This strategy has recently been harnessed with success to implement a CRISPR/Cas9 screening that unveiled the ability of BCL2 apoptosis regulator (BCL2) to endogenously suppress DC activation by preserving mitochondrial integrity and hence limiting the spontaneous release of interferogenic mitochondrial DNA species (Miller et al., 2021) in the cytosol (Kepp et al., 2023; Zhao et al., 2023). Of note, such a mitochondrial checkpoint appears to be a common mechanism that regulates not only adaptive immune responses elicited by DCs (Kepp et al., 2023; Zhao et al., 2023), but also cellular and organismal aging (Iske et al., 2020; Jiménez-Loygorri et al., 2024; Victorelli et al., 2023) as well as the ability of malignant cells to engage immune effectors upon treatment (Yamazaki et al., 2020), with an extra layer of negative regulation by apoptotic caspases (Han et al., 2020; Ning et al., 2019; Rodriguez-Ruiz et al., 2019; Rongvaux et al., 2014; White et al., 2014).

Importantly, iniDCs can also be harnessed to implement *in vitro* screenings for additional DC functions including phagocytic competence and cross-presentation proficiency, standing out as a convenient platform for the discovery of so-called immunogenic cell death (ICD) enhancers, i.e., molecules that improve the ability of dying cancer cells to elicit tumor-targeting immune responses by acting on DCs rather than malignant cells themselves (Galluzzi et al., 2024). Moreover, by offering access to a large and homogeneous amount of mature DCs, the iniDC platform enables a variety of *in vivo* assays, including prophylactic and therapeutic vaccination experiments in syngeneic mice optionally bearing specific genetic modifications or co-administrered with the pharmacological agent of choice (Galluzzi et al., 2024).

Perhaps the most important weakness of the iniDC platform is that (at least until now) it has only been implemented in mouse DCs. That said, while employing a Dox-inducible construct for the expression of the SV40 large T-antigen to conditionally immortalize human DC precursors (similar to the approach used to generated iniDCs) (Richter et al., 2013) would extend the *in vitro* screening potential offered by iniDCs to the human system, human DCs would remain incompatible with *in vivo* applications, with the partial (and hitherto underdeveloped) exception of humanized mice (Chuprin et al., 2023). Despite this (and potentially other) limitation(s), iniDCs stand out as preferential screening tools for the development of novel immunomodulatory agents with a broad range of applications.

Acknowledgments

We are indebted to Sebastian Thieme (University Clinic "Carl Gustav Carus" Dresden, Dresden, Germany) for the kind gift of iniDCs. O.K. is supported from Agence National de la Recherche (ANR) and Institut National du Cancer (INCa). G.K. is supported by the Ligue contre le Cancer (équipe labellisée); ANR—Projets blancs; AMMICa US23/CNRS UMS3655; Association pour la recherche sur le cancer (ARC); Cancéropôle Ile-de-France; Fondation pour la Recherche Médicale (FRM); a donation by Elior; Equipex Onco-Pheno-Screen; European Joint Programme on Rare Diseases (EJPRD); European Research Council Advanced Investigator Award (ERC-2021-ADG, ICD-Cancer, Grant No. 101052444), European Union Horizon 2020 Projects Oncobiome, Prevalung (Grant No. 101095604) and Crimson; INCa; Institut Universitaire de France; LabEx Immuno-Oncology ANR-18-IDEX-0001; a Cancer Research ASPIRE Award from the Mark Foundation; the RHU Immunolife; Seerave Foundation; SIRIC Stratified Oncology Cell DNA Repair and Tumor Immune Elimination (SOCRATE); and SIRIC Cancer Research and Personalized Medicine (CARPEM). This study contributes to the IdEx Université de Paris ANR-18-IDEX-0001. Views and opinions expressed are those of the author(s) only and do not necessarily reflect those of the European Union, the European Research Council or any other granting authority. Neither the European Union nor any other granting authority can be held responsible for them. The LG lab is/has been supported (as a PI unless otherwise indicated) by one NIH R01 grant (#CA271915), by two Breakthrough Level 2 grants from the US DoD BCRP (#BC180476P1, #BC210945), by a grant from the STARR Cancer Consortium (#I16-0064), by a Transformative Breast Cancer Consortium Grant from the US DoD BCRP (#W81XWH2120034, PI: Formenti), by a U54 grant from NIH/NCI (#CA274291, PI: Deasy, Formenti, Weichselbaum), by the 2019 Laura Ziskin Prize in Translational Research (#ZP-6177, PI: Formenti) from the Stand Up to Cancer (SU2C), by a Mantle Cell Lymphoma Research Initiative (MCL-RI, PI: Chen-Kiang) grant from the Leukemia and Lymphoma Society (LLS), by a Rapid Response Grant from the Functional Genomics Initiative (New York, US), by a pre-SPORE grant (PI: Demaria, Formenti), a Collaborative Research Initiative Grant and a Clinical Trials Innovation Grant from the Sandra and Edward Meyer Cancer Center (New York, US), by startup funds from the Department of Radiation Oncology at Weill Cornell Medicine (New York, US), by industrial collaborations with Lytix Biopharma (Oslo, Norway), Promontory (New York, US) and Onxeo (Paris, France), as well as by donations from Promontory (New York, US), the Luke Heller TECPR2 Foundation (Boston, US), Sotio a.s. (Prague, Czech Republic), Lytix Biopharma (Oslo, Norway), Onxeo (Paris, France), Ricerchiamo (Brescia, Italy), and Noxopharm (Chatswood, Australia).

Competing interests

O.K. is a scientific cofounder of Samsara Therapeutics. G.K. is on the Board of Directors of the Bristol Myers Squibb Foundation France. G.K. is a scientific cofounder of everImmune, Osasuna Therapeutics, Samsara Therapeutics and Therafast Bio. G.K. is in the scientific advisory boards of Hevolution, Institut Servier, Longevity Vision Funds and Rejuveron Life Sciences. G.K. is the inventor of patents covering therapeutic targeting of aging, cancer, cystic fibrosis and metabolic disorders. G.K.'s wife, Laurence Zitvogel, has held research contracts with Glaxo Smyth Kline, Incyte, Lytix, Kaleido, Innovate Pharma, Daiichi Sankyo, Pilege, Merus, Transgene, 9 m, Tusk and Roche, was on the Board of Directors of Transgene, is a cofounder of everImmune, and holds patents covering the treatment of cancer and the therapeutic manipulation of the microbiota. G.K.'s brother, Romano Kroemer, was an employee of Sanofi and now consults for Boehringer-Ingelheim. The funders had no role in the design of the study; in the writing of the manuscript, or in the decision to publish the results. LG is/has been holding research contracts with Lytix Biopharma, Promontory and Onxeo, has received consulting/advisory honoraria from Boehringer Ingelheim, AstraZeneca, OmniSEQ, Onxeo, The Longevity Labs, Inzen, Imvax, Sotio, Promontory, Noxopharm, EduCom, and the Luke Heller TECPR2 Foundation, and holds Promontory stock options. The other authors have no conflicts of interest to declare.

References

Blander, J. M. (2018). Regulation of the cell biology of antigen cross-presentation. *Annual Review of Immunology*, *36*, 717–753. https://doi.org/10.1146/annurev-immunol-041015-055523.

Brubaker, S. W., et al. (2015). Innate immune pattern recognition: A cell biological perspective. *Annual Review of Immunology*, *33*, 257–290. https://doi.org/10.1146/annurev-immunol-032414-112240.

Cabeza-Cabrerizo, M., et al. (2021). Dendritic cells revisited. *Annual Review of Immunology*, *39*, 131–166. https://doi.org/10.1146/annurev-immunol-061020-053707.

Cerrato, G., et al. (2020). Quantitative determination of phagocytosis by bone marrow-derived dendritic cells via imaging flow cytometry. *Methods in Enzymology*, *632*, 27–37. https://doi.org/10.1016/bs.mie.2019.07.021.

Chen, G. Y., & Nuñez, G. (2010). Sterile inflammation: Sensing and reacting to damage. *Nature Reviews. Immunology*, *10*, 826–837. https://doi.org/10.1038/nri2873.

Chuprin, J., et al. (2023). Humanized mouse models for immuno-oncology research. *Nature Reviews. Clinical Oncology*, *20*, 192–206. https://doi.org/10.1038/s41571-022-00721-2.

Cruz, F. M., et al. (2017). The biology and underlying mechanisms of cross-presentation of exogenous antigens on MHC-I molecules. *Annual Review of Immunology*, *35*, 149–176. https://doi.org/10.1146/annurev-immunol-041015-055254.

Cultrara, C., et al. (2023). A biologic-device combination product delivering tumor-derived antigens elicits immunogenic cell death-associated immune responses against glioblastoma. *Journal for Immunotherapy of Cancer*, *11*. https://doi.org/10.1136/jitc-2023-006880.

Deckers, J., et al. (2023). Engineering cytokine therapeutics. *Nature Reviews Bioengineering*, *1*, 286–303. https://doi.org/10.1038/s44222-023-00030-y.

Fucikova, J., et al. (2022). An autologous dendritic cell vaccine promotes anticancer immunity in patients with ovarian cancer with low mutational burden and cold tumors. *Clinical Cancer Research*, *28*, 3053–3065. https://doi.org/10.1158/1078-0432.Ccr-21-4413.

Galluzzi, L., et al. (2024). Targeting immunogenic cell stress and death for cancer therapy. *Nature Reviews. Drug Discovery* (in press).

García-Martínez, E., et al. (2018). Trial watch: Immunostimulation with recombinant cytokines for cancer therapy. *Oncoimmunology*, 7, e1433982. https://doi.org/10.1080/2162402x.2018.1433982.

Gong, T., et al. (2020). DAMP-sensing receptors in sterile inflammation and inflammatory diseases. *Nature Reviews. Immunology*, *20*, 95–112. https://doi.org/10.1038/s41577-019-0215-7.

Han, C., et al. (2020). Tumor cells suppress radiation-induced immunity by hijacking caspase 9 signaling. *Nature Immunology*, *21*, 546–554. https://doi.org/10.1038/s41590-020-0641-5.

Harari, A., et al. (2020). Antitumour dendritic cell vaccination in a priming and boosting approach. *Nature Reviews. Drug Discovery*, *19*, 635–652. https://doi.org/10.1038/s41573-020-0074-8.

Hennessy, E. J., et al. (2010). Targeting toll-like receptors: Emerging therapeutics? *Nature Reviews. Drug Discovery*, *9*, 293–307. https://doi.org/10.1038/nrd3203.

Heras-Murillo, I., et al. (2024). Dendritic cells as orchestrators of anticancer immunity and immunotherapy. *Nature Reviews. Clinical Oncology*, *21*, 257–277. https://doi.org/10.1038/s41571-024-00859-1.

Hildner, K., et al. (2008). Batf3 deficiency reveals a critical role for CD8alpha+ dendritic cells in cytotoxic T cell immunity. *Science*, *322*, 1097–1100. https://doi.org/10.1126/science.1164206.

Hong, E., & Dobrovolskaia, M. A. (2024). Detection of antigen presentation by murine bone marrow-derived dendritic cells after treatment with nanoparticles. *Methods in Molecular Biology*, *2789*, 161–169. https://doi.org/10.1007/978-1-0716-3786-9_17.

Iske, J., et al. (2020). Senolytics prevent mt-DNA-induced inflammation and promote the survival of aged organs following transplantation. *Nature Communications*, *11*, 4289. https://doi.org/10.1038/s41467-020-18039-x.

Isnard, S., et al. (2020). Monitoring antigen cross-presentation by human dendritic cells purified from peripheral blood. *Methods in Enzymology*, *635*, 283–305. https://doi.org/10.1016/bs.mie.2020.01.004.

Jhunjhunwala, S., et al. (2021). Antigen presentation in cancer: Insights into tumour immunogenicity and immune evasion. *Nature Reviews. Cancer*, *21*, 298–312. https://doi.org/10.1038/s41568-021-00339-z.

Jiménez-Loygorri, J. I., et al. (2024). Mitophagy curtails cytosolic mtDNA-dependent activation of cGAS/STING inflammation during aging. *Nature Communications*, *15*, 830. https://doi.org/10.1038/s41467-024-45044-1.

Johnson, P., et al. (2022). Conventional type 1 dendritic cells (cDC1) as cancer therapeutics: Challenges and opportunities. *Expert Opinion on Biological Therapy*, *22*, 465–472. https://doi.org/10.1080/14712598.2022.1994943.

Kepp, O., et al. (2023). A mitochondrial checkpoint to adaptive anticancer immunity. *Oncoimmunology*, *12*, 2271693. https://doi.org/10.1080/2162402x.2023.2271693.

Kroemer, G., et al. (2022). Immunogenic cell stress and death. *Nature Immunology*, *23*, 487–500. https://doi.org/10.1038/s41590-022-01132-2.

Kyrysyuk, O., & Wucherpfennig, K. W. (2023). Designing cancer immunotherapies that engage T cells and NK cells. *Annual Review of Immunology*, *41*, 17–38. https://doi.org/10.1146/annurev-immunol-101921-044122.

Le Naour, J., & Kroemer, G. (2023). Trial watch: Toll-like receptor ligands in cancer therapy. *Oncoimmunology*, *12*, 2180237. https://doi.org/10.1080/2162402x.2023.2180237.

Marchi, S., et al. (2023). Mitochondrial control of inflammation. *Nature Reviews. Immunology*, *23*, 159–173. https://doi.org/10.1038/s41577-022-00760-x.

Mastellos, D. C., et al. (2024). A guide to complement biology, pathology and therapeutic opportunity. *Nature Reviews. Immunology*, *24*, 118–141. https://doi.org/10.1038/s41577-023-00926-1.

Mempel, T. R., et al. (2024). How chemokines organize the tumour microenvironment. *Nature Reviews. Cancer*, *24*, 28–50. https://doi.org/10.1038/s41568-023-00635-w.

Miller, K. N., et al. (2021). Cytoplasmic DNA: Sources, sensing, and role in aging and disease. *Cell*, *184*, 5506–5526. https://doi.org/10.1016/j.cell.2021.09.034.

Ning, X., et al. (2019). Apoptotic caspases suppress type I interferon production via the cleavage of cGAS, MAVS, and IRF3. *Molecular Cell*, *74*, 19–31. e17 https://doi.org/10.1016/j.molcel.2019.02.013.

Palucka, K., & Banchereau, J. (2012). Cancer immunotherapy via dendritic cells. *Nature Reviews. Cancer*, *12*, 265–277. https://doi.org/10.1038/nrc3258.

Richter, C., et al. (2013). Generation of inducible immortalized dendritic cells with proper immune function in vitro and in vivo. *PLoS One*, *8*, e62621. https://doi.org/10.1371/journal.pone.0062621.

Rodriguez-Ruiz, M. E., et al. (2019). Apoptotic caspases inhibit abscopal responses to radiation and identify a new prognostic biomarker for breast cancer patients. *Oncoimmunology*, *8*, e1655964. https://doi.org/10.1080/2162402X.2019.1655964.

Rongvaux, A., et al. (2014). Apoptotic caspases prevent the induction of type I interferons by mitochondrial DNA. *Cell*, *159*, 1563–1577. https://doi.org/10.1016/j.cell.2014.11.037.

Theisen, D. J., et al. (2018). WDFY4 is required for cross-presentation in response to viral and tumor antigens. *Science*, *362*, 694–699. https://doi.org/10.1126/science.aat5030.

Vacchelli, E., et al. (2013a). Trial watch: Toll-like receptor agonists for cancer therapy. *Oncoimmunology*, *2*, e25238. https://doi.org/10.4161/onci.25238.

Vacchelli, E., et al. (2013b). Trial watch: Dendritic cell-based interventions for cancer therapy. *Oncoimmunology*, *2*, e25771. https://doi.org/10.4161/onci.25771.

Vanpouille-Box, C., et al. (2019). Pharmacological modulation of nucleic acid sensors—Therapeutic potential and persisting obstacles. *Nature Reviews. Drug Discovery*, *18*, 845–867. https://doi.org/10.1038/s41573-019-0043-2.

Victorelli, S., et al. (2023). Apoptotic stress causes mtDNA release during senescence and drives the SASP. *Nature*, *622*, 627–636. https://doi.org/10.1038/s41586-023-06621-4.

Wculek, S. K., et al. (2020). Dendritic cells in cancer immunology and immunotherapy. *Nature Reviews. Immunology*, *20*, 7–24. https://doi.org/10.1038/s41577-019-0210-z.

White, M. J., et al. (2014). Apoptotic caspases suppress mtDNA-induced STING-mediated type I IFN production. *Cell*, *159*, 1549–1562. https://doi.org/10.1016/j.cell.2014.11.036.

Woo, S. R., et al. (2015). Innate immune recognition of cancer. *Annual Review of Immunology*, *33*, 445–474. https://doi.org/10.1146/annurev-immunol-032414-112043.

Worbs, T., et al. (2017). Dendritic cell migration in health and disease. *Nature Reviews. Immunology*, *17*, 30–48. https://doi.org/10.1038/nri.2016.116.

Yamazaki, T., et al. (2020). Mitochondrial DNA drives abscopal responses to radiation that are inhibited by autophagy. *Nature Immunology*, *21*, 1160–1171. https://doi.org/10.1038/s41590-020-0751-0.

Yang, K., et al. (2023). Antigen presentation in cancer—Mechanisms and clinical implications for immunotherapy. *Nature Reviews. Clinical Oncology*, *20*, 604–623. https://doi.org/10.1038/s41571-023-00789-4.

Yatim, N., et al. (2017). Dying cells actively regulate adaptive immune responses. *Nature Reviews. Immunology*, *17*, 262–275. https://doi.org/10.1038/nri.2017.9.

Yin, X., et al. (2021). Dendritic cell regulation of T helper cells. *Annual Review of Immunology*, *39*, 759–790. https://doi.org/10.1146/annurev-immunol-101819-025146.

Zhao, L., et al. (2021). A genotype-phenotype screening system using conditionally immortalized immature dendritic cells. *STAR Protocols*, *2*, 100732. https://doi.org/10.1016/j.xpro.2021.100732.

Zhao, L., et al. (2023). BCL2 inhibition reveals a dendritic cell-specific immune checkpoint that controls tumor immunosurveillance. *Cancer Discovery*, *13*, 2448–2469. https://doi.org/10.1158/2159-8290.Cd-22-1338.

CHAPTER ELEVEN

Optimization and intra-assay validation of a multiparametric flow cytometric test for monitoring circulating TREGs

Iole Macchia*, Floriana Iacobone, and Francesca Urbani

Department of Oncology and Molecular Medicine, Istituto Superiore di Sanità (ISS), Rome, Italy
*Corresponding author: e-mail address: iole.macchia@iss.it

Contents

Abstract

Multiparametric flow cytometry (MFC) represents an essential tool for immune monitoring, and validation of MFC panels is a fundamental prerequisite in routine laboratory settings as well as for translational and clinical research purposes.

Regulatory T cells (TREGs) constitute a subset of CD4+ effector T cells that modulate the immune response in numerous settings, including autoimmune disease, allergy,

Methods in Cell Biology, Volume 189
ISSN 0091-679X
https://doi.org/10.1016/bs.mcb.2024.06.005

microbial infection, tumor immunity, transplantation, and more. These cells comprise a small fraction of total CD4+ T cells in human peripheral blood and mouse spleen. In oncology, TREG cells are highly relevant, as they are involved in the suppression of the anti-tumor response in many types of cancer, to the extent that the first immune checkpoint inhibitor approved for clinical use in humans was a monoclonal antibody directed against CTLA-4, a molecule functionally associated with TREGs. Due to all these factors, robust assays are mandatory to accurately determine TREG cell frequency and function.

Here, we describe the validation of an 8-color flow-cytometry protocol for TREG detection and analysis in a real-world laboratory scenario. The entire process includes the workflow plan and the standard operating procedure resembling each phase, from the panel design to the staining, acquisition, and analysis steps. Validation is planned to be performed in replicates on fresh whole blood samples derived from multiple healthy subjects. The analytical validity of the TREG cell assay is ensured by testing the intra-assay accuracy.

The detailed procedure for the entire process is accompanied by important troubleshooting suggestions and other useful tips.

1. Introduction

Multiparametric flow cytometry (MFC) is a powerful, sensitive, and quantitative analytical method that allows the identification of defined cell types within a mixed cell population, thanks to its unique ability to detect, at a single-cell level, multiple parameters simultaneously, by using several fluorochromes with different emission wavelength. MFC has been widely used to investigate the systemic immune profile of subjects, in translational and clinical studies.

Regulatory T cells (TREGs) are a subset of CD4+ T lymphocytes, with a role in regulating or suppressing other cells of the immune system; thus, they ensure the maintenance of immunological self-tolerance and immune homeostasis by preventing autoimmune responses against self-antigens (Levings, Allan, & Picc, 2006).

These cells can be produced within the normal thymus (natural TREGs) or outside it by differentiation of naïve T cells (adaptive TREGs). Natural TREGs (nTREGs) constitutively express the CD25 receptor (IL-2-receptor α-chain), required for their development and survival through IL-2 responsiveness (Niec & Waldmann, 2013), and the nuclear transcription factor Forkhead box P3 (FoxP3), which is necessary for their development and for maintaining their suppressive function; FoxP3 is also transiently expressed in activated non-suppressor T cells (Deng, Song, & Greene, 2019). Beyond FoxP3, the immunosuppressive functions of TREGs cells are also influenced by other factors, including the Cytotoxic T lymphocyte antigen-4 (CTLA-4) and the glucocorticoid-induced TNF-receptor (GITR) (Wegrzyn, Kedzierska, & Obojski, 2023).

TREGs have been shown to suppress most immune cell populations, including lymphocytes, various types of macrophages, dendritic and B cells. In particular, they are able to suppress activation, proliferation, and cytokine production of $CD4^+$ T cells and $CD8^+$ T cells (Okeke & Uzonna, 2019).

Immuno-monitoring TREGs is crucial as these cells are involved in a wide range of diseases, including cancer, autoimmune diseases, infectious diseases, transplantation, and allergy (Sakaguchi et al., 2020). Since this lymphocyte subpopulation is altered in number and/or function in numerous pathologies, their frequency may be considered a promising immunological biomarker.

Increased levels of TREGs have been reported at tumor sites, draining lymph nodes as well as in the peripheral blood, in many types of human cancer, including melanoma, breast, colon, lung, prostate, liver, ovarian cancer as well as hematological malignancies (Plitas & Rudensky, 2020). TREGs are generally considered a poor prognostic factor since they may suppress the immune response against tumor antigens (Liyanage et al., 2002). In fact, high levels of circulating TREGs in melanoma patients have been correlated with poor overall survival and disease-free survival (Heldager et al., 2024).

For this reason, several studies, have attempted to reduce or deplete the number of these cells by immune checkpoint inhibitor (ICI) therapy, to enhance tumor immunity and to control tumor growth (Takeuchi & Nishikawa, 2016; Wolf et al., 2003). Indeed, some ICIs, such as Ipilimumab and Tremelimumab, are monoclonal therapeutic antibodies directed against CD152 (CTLA-4) molecules, characteristic receptors of these cells. Therefore, TREG levels in peripheral blood of melanoma patients at the time of diagnosis may be considered a potential biomarker not only for prognosis and survival, but even for prediction of response to ICIs.

To date, proposed unique cell markers have failed to fulfill the criteria of being exclusively expressed on these cells, since TREGs represent a heterogeneous cell population with distinct phenotypes and functional profiles (Miyara et al., 2009). Standardization of TREG staining by flow cytometry is challenging, notably because of the intracellular location of the FoxP3 transcriptional factor.

Up to now, rigorous analysis of multiparametric flow cytometry data, using various sample preparations and optimization steps for TREG cell quantification, has highlighted certain surface and intracellular markers as essential for univocally identifying TREGs, which are often defined as CD3+CD4+CD25+FoxP3+CD127− cells (Biancotto, Dagur, Fuchs, Langweiler, & McCoy, 2012; Santegoets et al., 2015).

Additional markers use for nTREGs are the above mentioned CTLA-4 and GITR, although it should be noted that these markers are also expressed by other T-cell types (e.g., activated T cells), so they are not in themselves unequivocally diagnostic. However, the role of these markers on other T cells is not clearly defined. The expression of intracellular CTLA-4 and FoxP3 is positively correlated with surface expression of CD25 and inversely correlated with expression of the CD127 marker (IL-7 receptor) (Liu et al., 2006). Higher levels of intracellular FoxP3 and CTLA-4 and lower levels of surface CD127 consistently distinguish CD25hi cells from CD25int/low and CD25− cells (Sakaguchi et al., 2020).

Other phenotypic markers (CD45RA and CD39) can define more precisely subsets of TREGs, including the CD45RA-CD39+ memory TREGs with ATPase-dependent suppressive activity (Borsellino et al., 2007; Miyara et al., 2009). TREGs can also produce soluble messengers, endowed with a suppressive function, including TGF-β, IL-10 and adenosine.

Overall, the validation of a panel for the immunomonitoring of TREGs is necessary to obtain reliable and comparable results (Pitoiset et al., 2018).

The purpose of this protocol is to optimize and validate a flow cytometric multiparametric assay for evaluating TREG frequency and functional status in human fresh peripheral blood, adaptable for use across different clinical and physiological scenarios. Here, we describe the validation process of an 8-color panel for identification and enumeration of human naïve/memory TREG subsets, focusing on the CD45RA-CD39+ memory suppressive subpopulation, utilizing markers (such as CD45, CD3, CD4, CD25, CD127, FoxP3, CD45RA, CD39), selected in the light of previous consensus reports (Santegoets et al., 2015). This panel can be also considered as a backbone to be supplemented with further drop-in markers. Specifically, this protocol will provide guidelines for intra-assay validation. Full inter-assay validation is not feasible in ex vivo fresh blood samples due to the possible time-dependent alterations in the distribution of certain immune cell subsets, including TREGs.

Based on our previous experience in harmonizing a MFC panel aimed at studying memory status of T cell subpopulations (Macchia et al., 2020), the protocol presented below is the outcome of a real-life validation conducted in our laboratory.

2. Materials

2.1 Common disposables

Below are listed the common disposable items routinely used in our laboratory (Note 1).

- 3 mL K2-EDTA blood collection tubes (e.g., #367838 Vacutainer tubes, BD, NJ, USA).
- 96 U-bottom well plates (e.g., #3879 Corning, NY, USA).
- 5 mL round bottom polystyrene test tubes (e.g., #352054 Falcon Corning, NY, USA) (Note 2).
- 1.5 mL safe lock vials (e.g., #022363204 Eppendorf, DE, EUR).
- Pipet tips (e.g., # F161630, # F161930, #F161670 D10/D200/D1000 Diamond tips Gilson, WI, USA).
- 25-, 10-, 5-, 2-mL sterile disposable pipettes (e.g., #357535, #357551, #357543, #357507, Falcon Corning, NY, USA).

2.2 Reagents

Below are listed the reagents routinely used in our laboratory (Note 3).

- Deionized water.
- Dulbecco-Phosphate Buffered Saline (DPBS), w/o Ca and Mg (e.g., #17-512F Lonza, CH, EU) (Note 4).
- Fetal bovine serum (FBS) (e.g., #35-079 Corning, NY, USA) (Notes 5 and 6).
- EDTA concentrate (e.g., #E7889 MilliporeSigma-Merck, MO, USA) (Note 7).
- Sodium azide concentrate (e.g., #71290 MilliporeSigma-Merck, MO, USA) (Note 8).
- Compensation beads (e.g., #B22804 VersaComp Antibody Capture Bead Kit, Beckman Coulter, CA, USA) (Note 9).
- Lysing solution 10× (e.g., #349202 BD Biosciences, CA, USA) (Note 10).
- Fluorescent-conjugated anti-human antibodies (Note 11) as per Table 1.
- FoxP3/Transcription Factor Staining Buffer Set (# 00-5523-00 e-Biosciences, MA, USA), which includes fixation/permeabilization 4× buffer, permeabilization buffer 10× and diluent (Notes 12 and 13).
- Fc Receptor Blocking Solution (e.g., # 422302 Human TruStain FcX™, Biolegend, CA, USA) (Note 14).

Table 1 MFC panel composition.

Cell localization	Antigen/ligand	Fluorochrome	Clone	µL/sample	ID#	Manufacturer	Nationality (US)
Surface	CD4	FITC	RPA-T4	10	555346	BD Pharmingen	CA
Intranuclear	FoxP3 (Note 49)	PE	PCH101	5	12-4776-42	e-Bioscience	MA
Surface	CD3	PE Cy5.5	SP34-2	10	552852	BD Pharmingen	CA
Surface	CD25	PE Cy7	M-A251	5	557741	BD Pharmingen	CA
Surface	CD127	APC	HIL-7R-M21	10	558598	BD Pharmingen	CA
Surface	CD39	APC CY7	A1	5	328226	Biolegend	CA
Surface	CD45RA	Pacific Blue	HI100	5	304123	Biolegend	CA
Surface	CD45	Krome Orange	J.33	20	B36294	Beckman Coulter	CA

2.3 Equipment

Below are listed the equipment routinely used in our laboratory (Note 15).

- Class A biological safety cabinet (e.g., #8511 Kottermann, DE, EU) (Note 16).
- Centrifuge (e.g., #5810R Eppendorf, DE, EU) (Note 17).
- Vortex (e.g., # VWRI444-1372 VWR, PA, USA).
- Micropipette Set (e.g., #F167370 Pipetman L Starter kit, Gilson, WI, USA).
- Flow Cytometer (e.g., Gallios Beckman Coulter, CA, USA) (Note 18).
- Pipettor (e.g., #075002 Pipet-Aid Drummond, PA, USA).

2.4 Software

Below are listed the software routinely used in our laboratory (Note 19).

- Flow cytometry acquisition and analysis software (e.g., Kaluza V1.3, Beckman Coulter, CA, USA).
- Analysis software (e.g., Microsoft Office 2021 Excel or SPSS statistical processor, IBM-SPSS V28, IBM Corporate New York, NY, USA) (Note 20).

3. Methods

3.1 Safety warning

This protocol involves the use of human blood derivatives. Operators should have undergone appropriate training and must adhere to laboratory safety guidelines for blood derivative products, including the proper adoption of personal protective equipment (PPE) and controlled environment precautions (CEP).

3.2 Preliminary procedures

1. The project workflow is planned, as illustrated in Fig. 1A.
2. Sample processing is set up as well (Fig. 1B), by in-house adapting manufacturer instructions of the eBioscience FoxP3/Transcription Factor Staining Buffer Set (Note 21).

3.3 Panel design

1. A 8-color MFC panel in liquid format is designed (Table 1 in Section 2.2), based on a previous consensus report (Santegoets et al., 2015), by selecting antibody clones of the following markers:

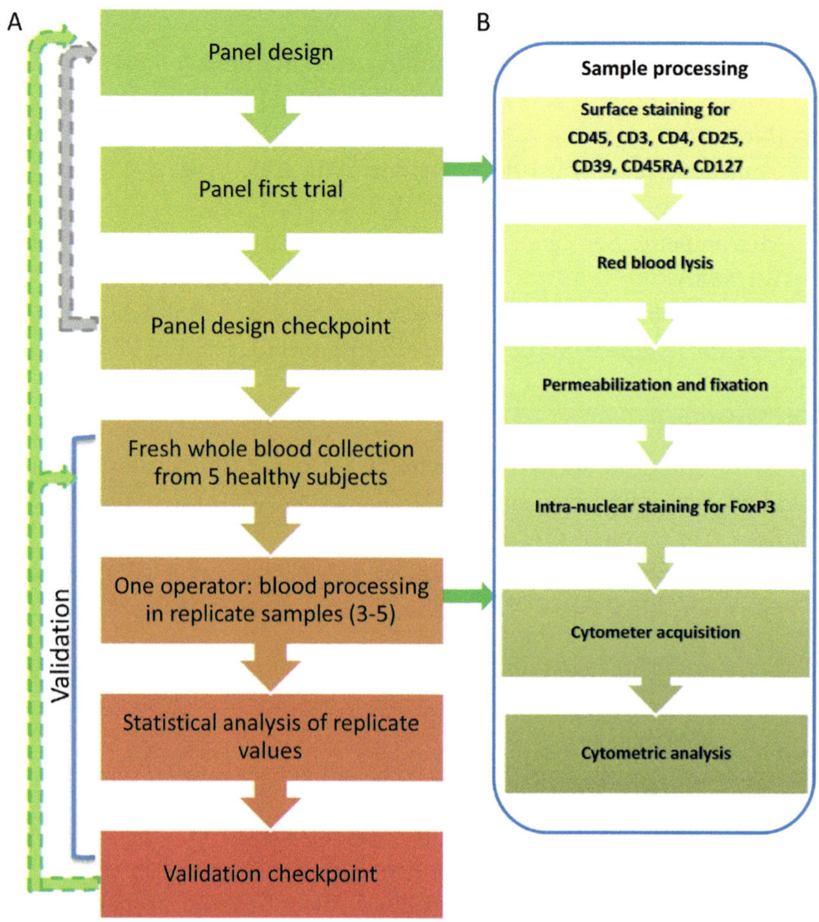

Fig. 1 Experimental workflow. (A) Flowchart of the validation of a flow cytometry panel for TREG monitoring in human fresh peripheral blood. Dotted arrows indicate potential re-design and re-validation steps. (B) Flow chart of sample staining, acquisition and analysis procedure.

- anti-CD45, -CD3, -CD4, -CD25, -CD127 and -FoxP3 conjugated antibodies, as minimally required markers.
- anti-CD45RA and -CD39, as additional markers.

2. Antibody-fluorochrome combinations are defined according to commonly recommended procedures for panel design (Flores-Montero et al., 2019) (Note 22).

3.4 Conjugated antibody titration

Each conjugated monoclonal antibody is titrated to define optimal working concentration. This procedure is strongly advised (Brando & Sommaruga, 1993), especially when constructing a multiparametric flow cytometry panel. The antibody titration procedure is not specifically addressed in this SOP, however, briefly, each antibody is tested at scalar dilutions in a twofold ratio starting from the dose recommended for the production batch (e.g., 10 μL, 5 μL, 2.5 μL, if manufacturer recommended dose is 10 μL). After acquisition, the optimal dose for each antibody is the one that shows the best resolution index (post-titration volumes are indicated in Table 1) (Notes 23 and 24).

3.5 Whole blood sample collection

1. Enrolled subjects sign written informed consent approved by the competent local ethics committee.
2. A 3 mL volume of whole blood is collected by venipuncture in K2-EDTA vacuum tubes from healthy volunteers (referred to as Subj#0, Subj#1 … #5) (Note 25). Subj#0 sample is dedicated to the first trial phase, while Subj#1 to #5 samples are dedicated to the validation phase.
3. Samples are left at room temperature until the beginning of the staining procedure, which is performed within 4 h from blood withdrawal.

3.6 Panel compensation set-up

The compensation matrix is established by VersaComp bead staining for each single pre-titrated conjugated antibody, according to manufacturer instructions (Note 26). To generate a compensation matrix, automatic procedure is applied and manual adjustments are operated, if necessary.

3.7 Panel first trial

This phase is performed to set up the correct PMT voltages and to verify compensation matrix as well as to check the overall good outcome of the panel.

3.7.1 Sample collection

1. Blood is collected from 1 enrolled subject (Subj#0), as described in Section 3.5.

2. 200 μL of Subj#0 whole blood are pipetted into the following 5 mL round bottom polystyrene tubes:
   - 1 unstained tube.
   - 1 fully stained (8-color) tube.

3.7.2 Sample staining

Warning: Flow cytometry procedure must be performed in the dark to preserve fluorochrome incidental excitation.

3.7.2.1 Working solutions (Note 27)

Working solutions are prepared and kept at RT until usage. Appropriate volumes are calculated for each sample to be processed, as follows:

- Staining solution: DPBS supplemented with 2% FBS, 2 mM EDTA, 0.09% Sodium Azide (Note 28).
- Lysing solution 1×: concentrated solution is diluted 1:10 in deionized water (Note 29).
- Fixation/Permeabilization 1×: 4× Fixation/Permeabilization buffer is diluted 1:4 in Diluent (Note 30).
- Permeabilization Solution 1×: 10× Permeabilization Buffer is diluted 1:10 in distilled water (Note 31).

3.7.2.2 Surface staining

1. A mix of pre-titrated surface antibodies (Table 1 included in Section 2.2), diluted in staining solution in a final volume of 50 μL for each sample, is prepared (Notes 32 and 33).
2. 50 μL of the surface antibody mixture are added to the 8-color tube(s).
3. Cells are gently mixed and incubated 15′ at RT in the dark.
4. To lyse red blood cells, 3 mL of 1× BD Lysing Solution are directly added to each tube.
5. Cells are mixed by inversion 2–3 times and incubated 10′ at RT in the dark (Note 34).
6. Cells are centrifuged at 500×*g* for 5′ at RT.
7. Supernatants are discarded.
8. 4 mL of staining buffer are added to each tube.
9. Cells are centrifuged at 500×*g* for 5′ at RT.
10. Supernatants are discarded.
11. Pellets are resuspended in the residual volume of staining buffer.

3.7.2.3 Fixation, permeabilization and intranuclear staining

From this point onwards, it is possible to transfer the contents of the tubes into the wells of a 96-well U-bottom plate (Note 35). This operation facilitates the management of a large number of samples. The volumes of the reagents relative to the plate procedure are indicated in parentheses.

1. 1 mL of freshly prepared fixation/permeabilization working solution 1× is added to each sample (250 μL) (see Section 3.7.2.1).
2. Cells are vortexed and incubated for 30′ at RT in the dark.
3. Cells are centrifuged at 800×*g* for 5′ at RT.
4. Supernatants are discarded (Note 36).
5. 2 mL of staining buffer are added to each sample (250 μL).
6. Cells are centrifuged at 800×*g* for 5′ at RT.
7. Supernatants are discarded (Note 37).
8. 1 mL of freshly prepared Permeabilization solution 1× (see Section 3.7.2.1) is added to each sample (250 μL).
9. Cells are centrifuged at 800×*g* for 5′ at RT.
10. Supernatants are discarded (Notes 38 and 39).
11. Cells are resuspended in the residual volume of Permeabilization Buffer 1×.
12. A dilution of freshly prepared anti-FoxP3 antibody is obtained in 1× permeabilization buffer (the appropriate pre-titrated volume of anti-FoxP3 is diluted in 50 μL of 1× permeabilization buffer for each sample) (Note 40).
13. 50 μL of diluted anti-FoxP3 are added to each sample and gently mixed.
14. Cells are incubated 30′ at RT in the dark.
15. 2 mL of staining buffer are added to each sample (250 μL).
16. Cells are centrifuged at 800×*g* for 5′ at RT.
17. Supernatants are discarded (Note 41).
18. Cells are resuspended in 300 μL of DPBS (Note 42).
19. Cells are left at 4 °C in the dark until acquisition (Note 43).

3.7.3 Sample acquisition

1. The unstained tube is run to set up morphological and fluorescent basal parameters.
2. The compensation matrix, previously calculated (see Section 3.6), is recalled. Manual adjustment can be performed, if necessary.
3. 250,000 total events (or 50,000 events in the Lymphocyte gate) are acquired for each 8-color-stained tube, with a flow rate <500 events/s for optimal resolution, recording each antibody fluorescence, FS-H, FS-A, SS-A and TIME channels (Note 44).

3.7.4 MFC analysis

1. Gating strategy (Fig. 2A) is performed by Kaluza software. For each 8-color sample, the following plots are drawn:

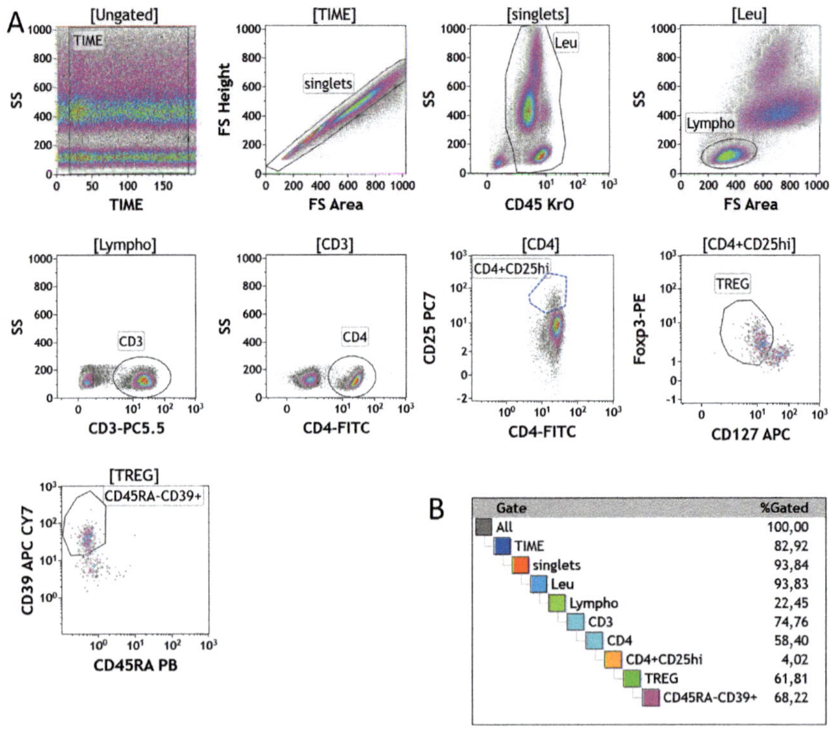

Fig. 2 Gating strategy and gate statistics for TREGs. (A) Representative dot plots from a healthy subject whole blood sample. TREGs are defined as CD3 + CD4 + CD25hiCD127-Foxp3+. Suppressive memory TREGs are defined as CD45RA-CD39+. (B) Representative gate statistics for whole blood samples.

- A TIME vs SS-A plot to select a stable acquisition time lapse by a gate (TIME).
- Within TIME: a FS-H vs FS-A plot to select SINGLET events.
- Within the singlet event gate: a CD45 vs FS-A plot to select LEUKOCYTES.
- Within the leukocyte gate: a FS-A vs SS-A plot to identify the LYMPHOCYTE population.
- Within the lymphocyte gate: a CD3 vs SS-A plot to identify CD3+ lymphocytes.
- Within the CD3 + lymphocyte gate: a CD4 vs SS-A plot to identify CD4+ lymphocytes.
- Within the CD4 + lymphocyte gate: a CD25+/CD4+ plot to identify CD4 + CD25high events.

- Within the CD4 + CD25high event gate: a CD127 vs FoxP3 plot, to identify CD127-FoxP3+ events, also known as *TREGs*.
- Within the TREG gate: a CD39/CD45RA plot, to identify CD45RA-CD39+ events, also described as *memory TREGs with ATPase-dependent suppressive activity*.

2. For each .fcs file, gate statistics are generated by Kaluza software (Fig. 2B).

3.7.5 First trial checkpoint

First trial is meant to assess the feasibility of compensation, fluorochrome compatibility, overall performance of the analysis, and to define the gating strategy design, as well. If the panel composition did not meet the requirements for proper acquisition and analysis (lack of compensation or other issues), the entire process is considered for repetition until a satisfactory result is achieved (Fig. 1A gray dotted arrow). The subsequent panel validation is launched when a satisfactory checkpoint result is achieved.

3.8 Panel validation

The entire procedure (blood collection, staining, acquisition) is conducted within the same day for each subject under study.

3.8.1 Sample collection

1. Blood is collected from 5 enrolled subjects (Subj#1 to #5), as described in Section 3.5.
2. 200 μL of each blood sample are aliquoted in three different 5 mL U-bottom polystyrene tubes (a, b, c) (Note 45).

3.8.2 Sample staining

Sample staining is performed as described in Section 3.7.2.

3.8.3 Sample acquisition

1. On the cytometer acquisition software, the PMT set up and compensation matrix derived from the first trial session are recalled. If needed, compensation matrix is manually adjusted before acquisition.
2. 250,000 total events (or 50,000 events in the lymphocyte gate) are acquired for each 8-color-stained tube, with a flow rate <500 events/s for optimal resolution, recording each antibody fluorescence, FS-H, FS-A, SS-A and TIME channels (Note 46).

3.8.4 MFC analysis

1. Each .fcs file is analyzed according to the gating strategy described in Section 3.7.4 and in Fig. 2A.
2. For each .fcs file, gate statistics is generated by Kaluza software (Fig. 2B).

3.8.5 Statistical analysis

1. Each .fcs file gate statistics are exported from Kaluza software into an Excel file or into a statistical processor database (Note 47).
2. To explore results, distribution of each parameter of interest by each subject analyzed is graphed (Fig. 3A).
3. Outliers are appropriately eliminated.
4. To estimate intra-assay precision, Coefficient of Variation (CV = SD/Mean) for each parameter of interest, related to each subject, is calculated. The median CV for each parameter and for each subject, is calculated as well (Fig. 3B). CV values below 0.20 are generally considered as optimal (Note 48).
5. A heatmap of calculated CVs is drawn for a quick overview of the results (Fig. 3B).
6. Observations are provided based on the obtained results, highlighting any parameters of interest that may be more prone to excessive variability.

3.8.6 Validation checkpoint

The opportunity to review the panel composition, staining method, gating strategy or final analysis should be considered in case one or more parameters of interest are found to be insufficiently accurate. The entire procedure should be repeated until satisfactory results are achieved (Fig. 1A, green dotted arrow).

4. Concluding remarks

The protocol described herein offers an easy and integrated workflow to validate a panel designed to assess frequency and activation status of circulating TREGs on a small volume of whole blood samples. The described panel may be adapted to the characterization of PBMC or tissue infiltrating TREGs and can also be further expanded by adding supplementary markers capable of qualifying the maturation and exhaustion state of TREGs cells even more finely (Manuszak, Brainard, Thrash, Hodi, & Severgnini, 2020).

A

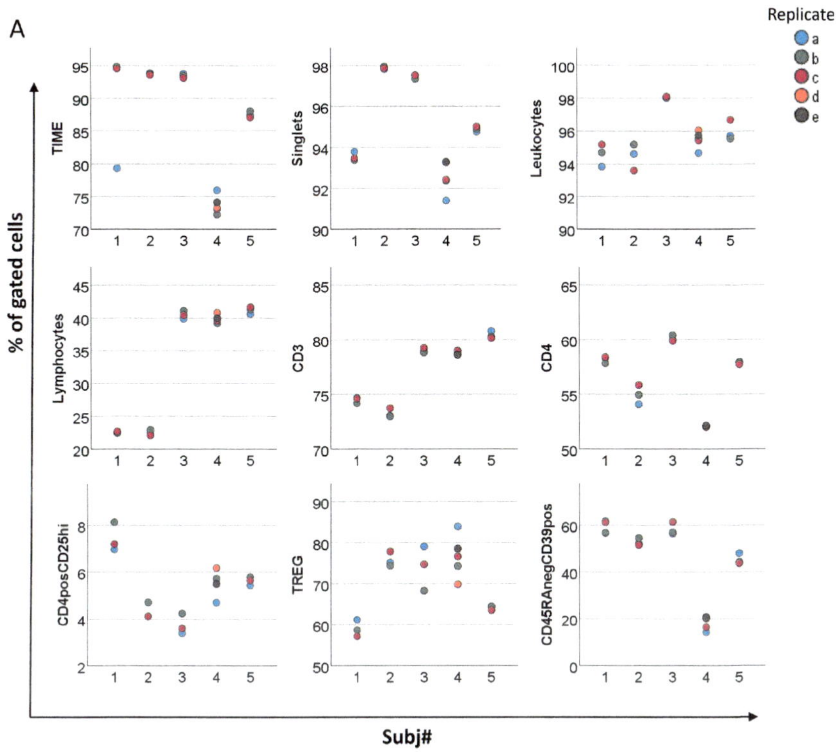

B

CV

	Subj#1	Subj#2	Subj#3	Subj#4	Subj#5	Parameter Mean
TIME	0,10	0,00	0,00	0,02	0,01	0,03
Singlets	0,00	0,00	0,00	0,01	0,00	0,00
Leukocytes	0,01	0,01	0,00	0,01	0,01	0,01
Lymphocytes	0,01	0,02	0,01	0,01	0,01	0,01
CD3	0,00	0,01	0,00	0,00	0,00	0,00
CD4	0,01	0,02	0,00	0,00	0,00	0,01
CD4posCD25hi	0,08	0,08	0,12	0,10	0,03	0,08
TREG	0,03	0,02	0,07	0,07	0,01	0,04
CD45RAnegCD39pos	0,05	0,03	0,05	0,16	0,05	0,07
Subject Mean	0,03	0,02	0,03	0,04	0,01	

C

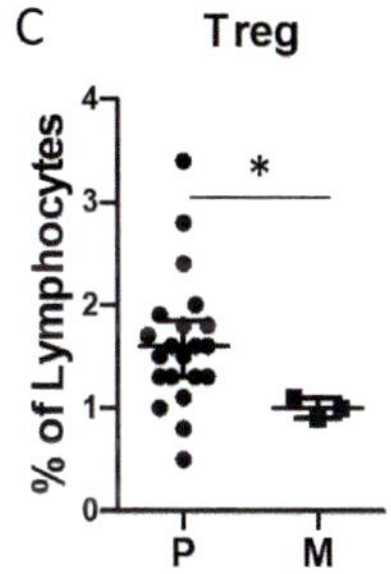

Fig. 3 Parameter distribution, CV heatmap and an applicative example. (A) Raw frequency distribution of subpopulation of interest. (B) Heatmap of CV values calculated for each parameter of interest on replicate results. (C) An applicative example: TREG frequency was assessed in primitive (P, $n = 21$) and metastatic (M, $n = 3$) breast cancer patients. *P calculated by U-Mann Whitney non-parametric test (unpublished data).

The described panel has passed intra-assay validation in our laboratory, and we have been applying it in several clinical and translational studies, such as in the context of breast cancer patients, where we found that primitive cancer patients had statistically significant higher levels of TREG cells if compared to metastatic patients (Fig. 3C, unpublished data).

A subsequent step could involve verifying its inter-laboratory reliability through a harmonization project, making it applicable in the context of immunological monitoring within multicenter studies (Macchia et al., 2020).

5. Notes

1. Catalog number and provider are indicated as a reference, but equivalent disposable items with equivalent characteristics may be used.
2. Check compatibility with your own flow cytometer.
3. Reagents with equivalent characteristics may be used, except for the indicated rat anti-human FoxP3 and the FoxP3/Transcription Factor Staining Buffer Set, which are strongly recommended.
4. Not classified under the Hazardous Products Current Regulations.
5. Heat-inactivated for 30′ at 56 °C.
6. This substance is classified as not hazardous according to the current regulation.
7. H332—Acute toxicity, Inhalation. H373—Specific target organ toxicity—repeated exposure, Respiratory Tract.
8. H300 + H310 + H330—Fatal if swallowed, in contact with skin or if inhaled. H373—May cause damage to organs (Brain) through prolonged or repeated exposure if swallowed. H410—Very toxic to aquatic life with long lasting effects.
9. Not classified as hazardous according to current regulation.
10. H302 + H312 + H332—Harmful if swallowed, in contact with skin or if inhaled. H314—Causes severe skin burns and eye damage. H317—May cause an allergic skin reaction. H335—May cause respiratory irritation. H341—Suspected of causing genetic defects. H350—May cause cancer. H370—Causes damage to organs. H373—May cause damage to organs through prolonged or repeated exposure. H402—Harmful to aquatic life.
11. The fluorescent-conjugated antibodies contain no substances which at their given concentration, are considered to be hazardous to health.
12. This reagent is strongly recommended since, in our laboratory, it has been compared with several other commercial kits, yielding the best results for the intra-nuclear staining with eBioscence anti-human FoxP3.
13. H317—May cause an allergic skin reaction. H341—Suspected of causing genetic defects. H350—May cause cancer. H370—Causes damage to organs.
14. This item is not a hazardous substance and does not contain hazardous ingredients or substances according to current regulation.

15. Equipment with equivalent characteristics may be used.
16. To be turned on at least 15 min prior to use to allow for airflow stabilization.
17. To be used at 500–800 × *g*.
18. At least 2 lasers: 488 and 630/640 nm. It must undergo an internal quality control of alignment (required), sensitivity and linearity (highly recommended).
19. Software with equivalent characteristics may be used.
20. Statistical processor is optional, since equivalent analysis and graphs can be obtained by Excel software.
21. https://www.thermofisher.com/it/en/home/references/protocols/cell-and-tissue-analysis/protocols/staining-intracellular-antigens-flow-cytometry.html
22. These procedures ensure proper matching between the markers under examination and the fluorophores conjugated to them, preventing bad staining artifacts such as excessive spillover or fluorochrome spread.
23. Antibody titration can indeed be conducted on the biological matrix of interest following relevant protocols. However, low-expressed markers might yield faint results. Therefore, it is advisable to utilize commercial ad hoc beads to mitigate this issue.
24. Resolution index = (xi − xo)/(square root of SDi squared + SDo squared), where "i" represents positive events and "o" represents negative events, "x" is the mean, and "SD" is the standard deviation.
25. Whole blood samples are taken by a specialized medical or nurse, as per current regulations.
26. Panel compensation can be conducted either on the biological matrix of interest or on VersaComp beads (https://www.beckmancoulter.com/wsrportal/techdocs?docname=B25652AA.pdf), or similar. However, low-expressed markers might yield faint results when detected on biological samples. Therefore, it is advisable to utilize commercial ad hoc beads to mitigate this issue.
27. An excess volume of each working solution should be calculated to avoid potential pipetting errors, when processing multiple samples.
28. Staining solution can be stored at +4 °C for 1 month.
29. Dilute the 10× concentrate 1:10 with room temperature (20–25 °C), deionized water. The prepared solution is stable for 1 month when stored in a glass or high-density polyethylene (HDPE) container at room temperature.
30. It is advisable to use a freshly prepared dilution.

31. As in the previous note.
32. Appropriate volume for each antibody to be diluted is established within the pre-titration phase described in Section 3.4.
33. If multiple samples are processed, it is recommended to prepare the mixture with a margin of excess to avoid issues related to potential pipetting errors.
34. Observe if the color of the liquid becomes transparent red, confirming that lysis of the red blood cells has occurred.
35. Avoid using adjacent wells to prevent cross-contamination.
36. In the case of using 96 U-bottom plates, supernatants can be discarded by inverting the plate, ensuring that the cellular pellet remains adherent to the plastic.
37. As in the previous note.
38. As in the previous note.
39. Please note that, after fixation/permeabilization, the cell pellet will become small and faint, thus possibly less visible.
40. See Notes 33 and 34.
41. See Note 37.
42. The staining of FoxP3 is sensitive to the fixation method. Cell fixation with paraformaldehyde is not recommended for FoxP3 detection (Timperi, Barnaba, & Piconese, 2017).
43. Acquisition must be performed within 2h from staining since a delay may result in altered expression of some parameters.
44. During acquisition preview, gates are adjusted within the FSC-H vs SSC-H plot, doublets are excluded by creating a combination of same-channel bivariate plots utilizing Area vs Height (i.e., FSC-A vs FSC-H). If necessary and your software allows it, adjust the biexponential scale. Adjust each detector so that all populations are clearly defined, and the events of interest are not pushed up against axes.
45. Subj#4 samples is tested in 5 replicates (a, b, c, d, e) in our real-life project (Fig. 3A).
46. An elevated number of events is necessary to assure statistical significance also for rare cell populations (Cossarizza et al., 2021).
47. See Note 21.
48. As per example in Fig. 3B, in our real-life validation study, all parameters of interest passed accuracy testing.
49. There are several commercial anti-human FoxP3 Abs that may give slightly different results. In our opinion, the PCH101 clone (eBioscience) gives the most reliable results in combination with CD127 (Law et al., 2010).

Acknowledgments

We would like to thank the enrolled subjects who have voluntarily donated blood, the Blood Bank of the Umberto I Polyclinic, the cytometry facility at Istituto Superiore di Sanità (FAST), as well as Mrs. Fabiola Diamanti, and Mr. Teodoro Squatriti for their technical support. We also thank Valentina La Sorsa for her valuable help in revising the article.

Competing interests

Each author declares no conflict of interest that could significantly influence the objectivity or integrity of the results presented in this scientific paper.

References

Biancotto, A., Dagur, P. K., Fuchs, J. C., Langweiler, M., & McCoy, J. P. (2012). OMIP-004: In-depth characterization of human T regulatory cells. *Cytometry. Part A: The Journal of the International Society for Analytical Cytology*, *81*(1), 15–16. https://doi.org/10.1002/cyto.a.21158.

Borsellino, G., Kleinewietfeld, M., Di Mitri, D., Sternjak, A., Diamantini, A., Giometto, R., et al. (2007). Expression of ectonucleotidase CD39 by Foxp3+ Treg cells: hydrolysis of extracellular ATP and immune suppression. *Blood*, *110*(4), 1225–1232. https://doi.org/10.1182/blood-2006-12-064527.

Brando, B., & Sommaruga, E. (1993). Nationwide quality control trial on lymphocyte immunophenotyping and flow cytometer performance in Italy. *Cytometry*, *14*(3), 294–306. https://doi.org/10.1002/cyto.990140310.

Cossarizza, A., Chang, H. D., Radbruch, A., Abrignani, S., Addo, R., Akdis, M., et al. (2021). Guidelines for the use of flow cytometry and cell sorting in immunological studies (third edition). *European Journal of Immunology*, *51*(12), 2708–3145. https://doi.org/10.1002/eji.202170126.

Deng, G., Song, X., & Greene, M. I. (2019). FoxP3 in T reg cell biology: A molecular and structural perspective. *Clinical and Experimental Immunology*, *199*, 255–262. https://doi.org/10.1111/cei.13357.

Flores-Montero, J., Kalina, T., Corral-Mateos, A., Sanoja-Flores, L., Pérez-Andrés, M., Martin-Ayuso, M., et al. (2019). Fluorochrome choices for multi-color flow cytometry. *Journal of Immunological Methods*. https://doi.org/10.1016/j.jim.2019.06.009.

Heldager, N., Bjerregaard, H., Persson, G., Bojesen, S., Vauvert, T., & Hviid, F. (2024). Current research in immunology an increase in regulatory T cells in peripheral blood correlates with an adverse prognosis for malignant melanoma patients—A study of T cells and natural killer cells. *Current Research in Immunology*, *5*(September 2023), 100074. https://doi.org/10.1016/j.crimmu.2023.100074.

Law, J. P., Hirschkorn, D. F., Owen, R. E., Biswas, H. H., Philip, J., & Lanteri, M. C. (2010). The importance of Foxp3 antibody and fixation/permeabilization buffer combinations in identifying CD4+ CD25+ Foxp3+ regulatory T cells. *Cytometry A*, *75*(12), 1040–1050. https://doi.org/10.1002/cyto.a.20815.

Levings, M. K., Allan, S., & Picc, A. (2006). Functional dynamics of naturally occurring regulatory T cells in health and autoimmunity. *Advances in Immunology*, *92*(2006), 119–155. https://doi.org/10.1016/S0065-2776(06)92003-3.

Liu, W., Putnam, A. L., Xu-yu, Z., Szot, G. L., Lee, M. R., Zhu, S., et al. (2006). CD127 expression inversely correlates with FoxP3 and suppressive function of human CD4 + T reg cells. *The Journal of Experimental Medicine*, *203*(7), 1701–1711. https://doi.org/10.1084/jem.20060772.

Liyanage, U. K., Moore, T. T., Joo, H.-G., Tanaka, Y., Herrmann, V., Doherty, G., et al. (2002). Prevalence of regulatory T cells is increased in peripheral blood and tumor microenvironment of patients with pancreas or breast adenocarcinoma. *The Journal of Immunology*. https://doi.org/10.4049/jimmunol.169.5.2756.

Macchia, I., La Sorsa, V., Ruspantini, I., Sanchez, M., Tirelli, V., Carollo, M., et al. (2020). Multicentre harmonisation of a six-colour flow cytometry panel for Naïve/memory T cell Immunomonitoring. *Journal of Immunology Research, 2020*, 1–15. https://doi.org/10.1155/2020/1938704.

Manuszak, C., Brainard, M., Thrash, E., Hodi, F. S., & Severgnini, M. (2020). Standardized 11-color flow cytometry panel for the functional phenotyping of human T regulatory cells. *Journal of Biological Methods*, 7(2), 1–6. https://doi.org/10.14440/jbm.2020.325.

Miyara, M., Yoshioka, Y., Kitoh, A., Shima, T., Wing, K., Niwa, A., et al. (2009). Functional delineation and differentiation dynamics of human CD4 + T cells expressing the FoxP3 transcription factor. *Immunity, 30*, 899–911. https://doi.org/10.1016/j.immuni.2009.03.019.

Niec, R. E., & Waldmann, H. (2013). The plasticity and stability of regulatory T cells. *Nature Reviews Immunology, 13*(6), 461–467. https://doi.org/10.1038/nri3464.

Okeke, E. B., & Uzonna, J. E. (2019). The pivotal role of regulatory T cells in the regulation of innate immune cells. *Frontiers in Immunology, 10*(April), 1–12. https://doi.org/10.3389/fimmu.2019.00680.

Pitoiset, F., Barbié, M., Monneret, G., Braudeau, C., Pochard, P., Pellegrin, I., et al. (2018). A standardized flow cytometry procedure for the monitoring of regulatory T cells in clinical trials. *Cytometry Part B - Clinical Cytometry, 94*(5), 621–626. https://doi.org/10.1002/cyto.b.21622.

Plitas, G., & Rudensky, A. Y. (2020). Regulatory T cells in cancer. *Annual Review of Cancer Biology*. https://doi.org/10.1146/annurev-cancerbio-030419-033428.

Sakaguchi, S., Mikami, N., Wing, J. B., Tanaka, A., Ichiyama, K., & Ohkura, N. (2020). Regulatory T cells and human disease. *Annual Review of Immunology, 38*, 541–566. https://doi.org/10.1146/annurev-immunol-042718-041717.

Santegoets, S. J. A. M., Dijkgraaf, E. M., Battaglia, A., Beckhove, P., Britten, C. M., Gallimore, A., et al. (2015). Monitoring regulatory T cells in clinical samples: Consensus on an essential marker set and gating strategy for regulatory T cell analysis by flow cytometry. *Cancer Immunology, Immunotherapy, 64*(10), 1271–1286. https://doi.org/10.1007/s00262-015-1729-x.

Takeuchi, Y., & Nishikawa, H. (2016). Roles of regulatory T cells in cancer immunity. *International Immunology, 28*(8), 401–409. https://doi.org/10.1093/intimm/dxw025.

Timperi, E., Barnaba, V., & Piconese, S. (2017). Phenotypic and functional analysis of the suppressive function of human regulatory T cells. *Methods in Molecular Biology, 1514*, 139–151. https://doi.org/10.1007/978-1-4939-6548-9_12/FIGURES/2.

Wegrzyn, A. S., Kedzierska, A. E., & Obojski, A. (2023). Identification and classification of distinct surface markers of T regulatory cells. *Frontiers in Immunology, 13*(January), 1–12. https://doi.org/10.3389/fimmu.2022.1055805.

Wolf, A. M., Wolf, D., Steurer, M., Gastl, G., Gunsilius, E., & Grubeck-Loebenstein, B. (2003). Increase of regulatory T cells in the peripheral blood of cancer patients. *Clinical Cancer Research*.

CHAPTER TWELVE

Cytofluorometric assessment of calreticulin exposure on CD38$^+$ plasma cells from the human bone marrow

Manuel Beltrán-Visiedo[a], Alfonso Serrano-Del Valle[b], Nelia Jiménez-Aldúan[b], Ruth Soler-Agesta[a,b], Javier Naval[b], Lorenzo Galluzzi[a,c,d,*,†], and Isabel Marzo[b,*,†]

[a]Department of Radiation Oncology, Weill Cornell Medical College, New York, NY, United States
[b]Apoptosis, Immunity & Cancer Group, IIS Aragón, University of Zaragoza, Zaragoza, Spain
[c]Sandra and Edward Meyer Cancer Center, New York, NY, United States
[d]Caryl and Israel Englander Institute for Precision Medicine, New York, NY, United States
*Corresponding authors: e-mail address: deadoc80@gmail.com; imarzo@unizar.es

Contents

† Co-senior authors.

Methods in Cell Biology, Volume 189
ISSN 0091-679X
https://doi.org/10.1016/bs.mcb.2024.05.009

Abstract

Exposure of the endoplasmic reticulum chaperone calreticulin (CALR) on the surface of stressed and dying cells is paramount for their effective engulfment by professional antigen-presenting cells such as dendritic cells (DCs). Importantly, this is required (but not sufficient) for DCs to initiate an adaptive immune response that culminates with an effector phase as well as with the establishment of immunological memory. Conversely, the early exposure of phosphatidylserine (PS) on the outer layer of the plasma membrane is generally associated with the rapid engulfment of stressed and dying cells by tolerogenic macrophages. Supporting the clinical relevance of the CALR exposure pathway, the spontaneous or therapy-driven translocation of CALR to the surface of malignant cells, as well as intracellular biomarkers thereof, have been associated with improved disease outcome in patients affected by a variety of neoplasms, with the notable exception of multiple myeloma (MM). Here, we describe an optimized protocol for the flow cytometry-assisted quantification of surface-exposed CALR and PS on $CD38^+$ plasma cells from the bone marrow of patients with MM. With some variations, we expect this method to be straightforwardly adaptable to the detection of CALR and PS on the surface of cancer cells isolated from patients with neoplasms other than MM.

1. Introduction

Most often, mammalian cells die in an immunologically silent or even tolerogenic manner, especially (1) when cell death is part of a purely physiological program of development or adult tissue homeostasis (Fuchs & Steller, 2015), and (2) when cell death is unrelated to bacterial or viral infection (Kroemer et al., 2022). From an organismal perspective, this is particularly important to avoid unwarranted autoreactive/autoimmune conditions that may otherwise emerge from physiological tissue turnover or from pathological conditions that cause cell death in the absence of infection (*e.g.*, ischemia, trauma) (Kroemer et al., 2022). In line with this notion, several mechanisms are in place to ensure that infection-unrelated cell death in mammals, which most often is precipitated by the apoptotic machinery (Czabotar & Garcia-Saez, 2023; Diepstraten et al., 2022; Vitale et al., 2023), goes unnoticed by the immune system. For instance, apoptotic caspases such as caspase 3 (CASP3) proficiently shut down inflammatory reactions as potentially elicited by mitochondrial outer membrane permeabilization (MOMP) and the consequent release of interferogenic mitochondrial DNA (mtDNA) in the cytosol (Giampazolias et al., 2017; Han et al., 2020; Ning et al., 2019; Rodriguez-Ruiz et al., 2019; Rongvaux et al., 2014; White et al., 2014; Yamazaki et al., 2020). Moreover CASP3 effectively stimulates the secretion of immunosuppressive molecules such as prostaglandin E_2 (PGE_2) (Böttcher et al., 2018; Huang et al., 2011; Zelenay et al., 2015), as well as the exposure of phosphatidylserine (PS) on the outer leaflet of the plasma membrane, which elicits the rapid and immunologically silent clearance of dying cells by macrophages (Boada-Romero et al., 2020; Doran et al., 2020; Fadok et al., 1992; Mehrotra & Ravichandran, 2022; Rothlin et al., 2021). Reflecting the critical importance of an efficient and immunologically silent removal of dying cells and their corpses by macrophages, *Jmjd6*$^{-/-}$ mice (which lack the receptor for surface-exposed PS) die at birth owing to the accumulation of dead cells in the brain and lung (Fadok et al., 2000; Li et al., 2003).

In pathological settings, the death of mammalian cells can also promote inflammation potentially coupled with adaptive immune reactions (Galluzzi et al., 2023). Most often, cell death drives inflammation in the absence of adaptive immunity when (1) cell death is not caused by infectious agents, and (2) the amount of dying cells, which release a number of adjuvant-like molecules cumulatively known as damage-associated molecular patterns (DAMPs) overwhelms the phagocytic capacity of macrophages (Galluzzi et al., 2020a). Conversely, adaptive immune responses culminating with a cytotoxic T lymphocyte (CTL)-mediated effector phase and the establishment of immunological memory can be initiated by dying cells when (1) they express endogenous antigenic determinants that are not covered by central tolerance (antigenicity), (2) cell death is associated with the spatiotemporally coordinated release of key DAMPs and cytokines that recruit and license professional antigen-presenting cells (APCs) such as dendritic cells (DCs) (adjuvanticity), and (3) the microenvironment of dying cells allows for the recruitment of DC precursors, their activation and their migration to secondary or tertiary lymphoid organs for antigen presentation to $CD4^+$ and $CD8^+$ T cells (microenvironmental permissiveness) (Galluzzi et al., 2020a; Gong et al., 2020; Kroemer et al., 2022; Krysko et al., 2012; Matzinger, 2002; Mellman et al., 2023). Interestingly, such an immunogenic form of cell death is not only driven by pathogens, but also by a variety of chemical and physical stimuli, provided that the antigenicity criteria is met (Jhunjhunwala et al., 2021; Lang et al., 2022; Stern et al., 2024). These include (but not limited to): specific chemotherapeutics (Galluzzi et al., 2020b; Sprooten et al., 2023), radiation therapy (Cytlak et al., 2022; McLaughlin et al., 2020; Rodriguez-Ruiz et al., 2020), oncolytic agents (Harrington et al., 2019; Kepp et al., 2020; Palanivelu et al., 2022; Shalhout et al., 2023), photodynamic therapy (Adkins et al., 2014; Garg et al., 2016), and extracorporeal photopheresis (Tatsuno et al., 2019;

Ventura et al., 2018). Interestingly, while apoptotic caspases have been shown to limit the immunogenicity of cell death by a number of mechanisms, both apoptotic and non-apoptotic instances of immunogenic cell death (ICD) have been reported (Casares et al., 2005; Efimova et al., 2020; Hadian & Stockwell, 2023; Lei et al., 2022; Yatim et al., 2015; Yuan & Ofengeim, 2023).

Calreticulin (CALR) is an endoplasmic reticulum (ER) chaperone that exerts prominent roles in protein folding, Ca^{2+} homeostasis and MHC Class I antigen presentation (Fucikova et al., 2021). Importantly, stressed and dying cells can expose CALR on their surface (Obeid et al., 2007), where it delivers potent pro-phagocytic signals to a variety of myeloid cells (including DC precursors) that express low-density lipoprotein receptor-related protein 1 (LRP1, also known as CD91) and possibly other scavenger receptors (Gardai et al., 2005; Ogden et al., 2001; Pawaria & Binder, 2011; Vandivier et al., 2002). Mechanistically, the ICD-associated CALR exposure is elicited in the context of the so-called "integrated stress response" (ISR), which is demarcated by the inactivating phoeukaryotic translation initiation factor 2 subunit alpha (EIF2S1, best known as eIF2α) (Bezu et al., 2018a, 2018b; Panaretakis et al., 2009).

Supporting the clinical relevance of this pathways in oncological settings, the level of surface-exposed CALR (as well as the degree of eIF2α phosphorylation) has been associated with improved clinical outcome in several cohorts of patients with cancer (Fucikova et al., 2015). Specifically, the spontaneous or therapy-driven exposure of CALR on the cancer cell surface has been linked with superior disease outcome in cohorts of patients with non-small cell lung carcinoma (NSCLC) (Fucikova et al., 2016a), acute myeloid leukemia (AML) (Fucikova et al., 2016b; Truxova et al., 2020; Wemeau et al., 2010), neuroblastoma (Hsu et al., 2005), colorectal carcinoma (Peng et al., 2010), as well as endometrial and ovarian cancer (Kasikova et al., 2019; Xu et al., 2018). That said, elevated levels of surface-exposed CALR have also been associated with poor clinical outcome in patients with multiple myeloma (MM) (Serrano Del Valle et al., 2022).

Here, we provide a simple cytofluorometric protocol to quantify CALR and PS exposure on the membrane of $CD38^+$ plasma cells isolated the bone marrow (BM) of patients with MM. With some variations, this method should be straightforwardly adaptable to the assessment of surface-exposed CALR and PS on malignant cells from patients with a variety of tumors.

2. Materials

2.1 Common disposables (*see Note 1*)

- Acrodisc® Syringe Filters for Prefiltration, 0.22 μM (Cytiva, #17194361)
- Falcon™ 5 mL Round-Bottom Polypropylene Test Tubes Without Cap (Fisher Scientific, #10519901)
- Nunc™ 15 mL Conical Sterile Polypropylene Centrifuge Tubes (Thermo Scientific™, #339650)
- Nunc™ 50 mL Conical Sterile Polypropylene Centrifuge Tubes (Thermo Scientific™, #339652)

2.2 Cells and reagents (*see Note 1*)

- 7-Actinomycin D (7-AAD) Viability Staining Solution (Biolegend, #420403)
- Annexin V-DY-634 PI Apoptosis Staining/Detection Kit (abcam, #ab214484) (*see Note 2*)
- BD Pharmingen™ APC Mouse Anti-Human CD38 (HIT2) (BD Pharmingen™, #560980)
- BD Pharmingen™ FITC Mouse Anti-Human CD38 (HIT2) (BD Pharmingen™, #560982)
- Bone marrow (BM) samples from patients with MM or other plasma cell malignancies (*see Notes 3–4*)
- Calreticulin monoclonal antibody (FMC 75), DyLight™ 488 (Enzo Life Sciences, #ADI-SPA-601-488-F)
- Fetal bovine serum (FBS), Sigma-Aldrich, #F7524)
- Gibco™ RPMI 1640 Medium, GlutaMAX™ Supplement (Gibco, #12027599)
- Histopaque®-1077 (Sigma-Aldrich, #10771)
- Mouse IgG1 Isotype Control (MOPC-21), DyLight™ 488 (Invitrogen, #MA1–191-D488)
- PBS—Phosphate-Buffered Saline (10×) pH 7.4, RNase-free (Thermo Scientific™, #AM9624) (*see Note 5*)
- Trypan Blue solution (Sigma-Aldrich, #T8154)

2.3 Equipment (*see Note 1*)

- Benchtop cell counter, such as Countess II automated cell counter (Invitrogen) (*see Note 6*)

- Benchtop centrifuge, such as Allegra® X-15R Centrifuge (Beckman Coulter)
- Class II vertical laminar air flow cabinet, such as Bio II Advance Plus (Telstar)
- Flow cytometer equipped with lasers for acquisition of fluorescence from FITC and DyLight™ 488 (488 nm, 610/20 nm filter), 7-AAD (488 nm, 690/50 filter) and DY-634 (640 nm, 710/50 nm filter), such as BD FACSCalibur™ (BD Biosciences)
- Humidified cell culture incubator, such as Heraeus Heracell 150i (Thermo Scientific™) (*see Note 7*)

2.4 Software (*see Note 1*)

- FlowJo™ v. 7.6.1 (FlowJo LLC)
- Prism v. 10.0.1 (GraphPad)

3. Methods

3.1 Culture and isolation of mononuclear cells from human BM samples

3.1.1 Medium preparation

1. Heat FBS at 56 °C for 30 min in a water bath, and centrifuge at 560 × *g* for 15 min (*see Note 8*).
2. Filter supernatant with a sterile 0.22 μm syringe filter and collect in a sterile 50 mL conical tube.
3. Generate complete bone marrow mononuclear cell (BMMC) medium by complementing RPMI 1640 Medium, GlutaMAX™ Supplement with 10% (v:v) heat-inactivated FBS.

3.1.2 BMMC isolation

1. Add 3–4 mL Histopaque®-1077 (according to the volume of BM sample) (*see Note 9*) to sterile 15 mL conical tubes.
2. Dilute BM samples 1:2 (v:v) with RPMI 1640 Medium, GlutaMAX™ Supplement (without heat-inactivated FBS) or PBS.
3. Gently layer BM samples onto Histopaque®-1077 by ensuring than phases remain separated (*see Notes 10,11*).
4. Centrifuge at 500 × *g* for 20 min while deactivating the rotor brake (*see Note 12*).
5. Remove the upper plasma layer with a P1000 micropipette or a sterile Pasteur pipette (*see Note 13*).

6. Collect BMMCs (which are located at the interphase between the plasma and Histopaque®-1077 phase) with a P1000 micropipette and transfer them into a sterile 15 mL conical tube containing 10 mL RPMI 1640 Medium, GlutaMAX™ Supplement.
7. Thoroughly wash BMMCs twice with RPMI 1640 Medium, GlutaMAX™ Supplement to eliminate traces of Histopaque®-1077 by centrifuging at 335 × *g* for 5 min and discarding the supernatant at each washing cycle.
8. Resuspend BMMCs in 10 mL BMMC complete medium.
9. Determine BMMC density by mixing 30 μL Trypan Blue solution and 30 μL BMMC suspension and counting with the Countess II automated cell counter (*see Note 6*).

3.2 Surface-expose CALR quantification

1. Transfer 5×10^5 primary BMMCs to 5 mL Round-Bottom Polypropylene Test Tubes (*see Note 14*), and centrifuge at 335 × *g* 5 for 5 min.
2. Resuspend BMMCs in 100 μL PBS complemented with 5% FBS plus 2 μL Calreticulin monoclonal antibody (FMC 75), DyLight™ 488 or Mouse IgG1 Isotype Control (MOPC-21), DyLight™ 488, 8 μL APC Mouse Anti-Human CD38 (HIT2), and 1 μL 7-AAD Viability Staining Solution.
3. Incubate at 4 °C for 30 min in the darkness (*see Note 15*).
4. Centrifuge BMMCs at 335 × *g* for 5 min and resuspend them in 200–300 μL PBS for acquisition of fluorescent signals by flow cytometry.

3.3 Acquisition and gating

1. Flow cytometry is performed on an instrument equipped with 488 and 635 nm lasers. For data acquisition, the following channels are used: SSC-H, FSC-H, FL1-H (488 nm laser, 530/30 filter), FL3-H (488 nm laser, 650LP filter) and FL4-H (635 nm laser, 661/16 filter).
2. Set laser voltage (gain) and compensation unstained cells and cells stained with each fluorochrome separately.
3. If possible, acquire >10.000 CD38$^+$ events per sample, ideally at a flow rate < 500 events/s (*see Note 16*).
4. For gating (Fig. 1), proceed as follows:
   - **a.** Gate CD38$^+$ cells on a SSC-H *vs.* CD38-APC (FL4-H) dot plot window (*see Note 17*).
   - **b.** Gate viable CD38$^+$/7-AAD$^-$ plasma cells on a CD38-APC (FL4-H) *vs.* 7-AAD (FL3-H) dot plot (*see Notes 18,19*).

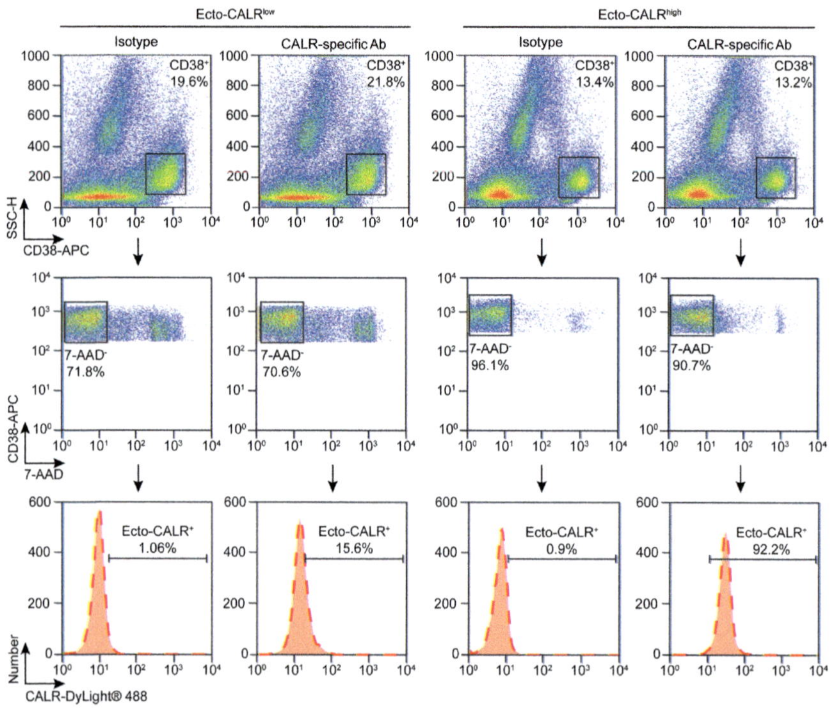

Fig. 1 Analysis of CALR exposure on live CD38$^+$ plasma cells from the human bone marrow. Bone marrow mononuclear cells (BMMCs) isolated from patient-derived bone marrow samples as detailed in Section 3.1 were stained with fluorescent antibodies specific for calreticulin (CALR), or isotype controls, and CD38 in the presence of 7-AAD as a vital dye, as detailed in Section 3.2, followed by signal acquisition on flow cytometry and gating as detailed in Section 3.3. Representative dot plots and histograms from a sample with low surface-exposed CALR (Ecto-CALR^{low}) and an Ecto-CALR^{high} sample are reported. Percentages indicate the number of events within specific gates.

c. Based on the CD38$^+$/7-AAD$^-$ dot plot, create a histogram for DyLight™ 488 (FL1-H). Set the gate for surface-expose CALR positivity based on the fluorescent signal of Mouse IgG1 Isotype Control (MOPC-21), DyLight™ (<1% positivity) (*see Note 20*).

3.4 Surface-exposed PS quantification

1. Transfer 5×10^5 primary BMMCs to 5 mL Round-Bottom Polypropylene Test Tubes (*see Note 14*), and centrifuge at $335 \times g$ 5 for 5 min.
2. Immediately resuspend BMMCs in 100 μL in Annexin Binding Buffer (ABB) as provided in the Annexin V-DY-634 PI Apoptosis Staining/Detection Kit (*see Note 21*).

3. Add 4 μL FITC Mouse Anti-Human CD38 (HIT2), 0.5 μL Annexin V-DY-634 (1 mg/mL) and 1 μL 7-AAD Viability Staining Solution to each sample (*see Note 22*).
4. Incubate at RT for 15–20 min.
5. Dilute samples with 200–300 μL ABB for acquisition of fluorescent signals by flow cytometry.

3.5 Acquisition and gating

1. Flow cytometry is performed on an instrument equipped with 488 and 635 nm lasers. For data acquisition, the following channels are used: SSC, FSC, FL1-H (488 nm laser, 530/30 filter), FL3-H (488 nm laser, 650LP filter), and FL4-H (635 nm laser, 661/16 filter).
2. For gating (Fig. 2), proceeds as follow:
   a. Gate CD38^{+} cells on a SSC-H *vs.* CD38-FITC (FL1-H) dot plot (*see Note 17*).
   b. Gate viable CD38^{+}/7-AAD^{-} plasma cells on a CD38-APC (FL4-H) *vs.* 7-AAD (FL3-H) dot plot (*see Note 18, 19*).
   c. Based on the CD38^{+}/7-AAD^{-} dot plot, create a histogram for DY-634 (FL4-H). Set the gate for surface-expose PS positivity based in the closed proximity of the major peak (*see Note 23*).

4. Notes

1. Virtually equivalent disposable, reagents, instruments and software can be purchased from multiple other providers at a similar cost.

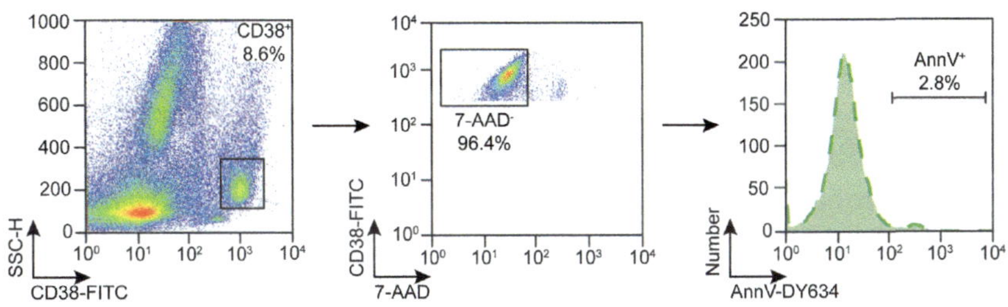

Fig. 2 Analysis of PS exposure on live CD38^{+} plasma cells from the human bone marrow. Bone marrow mononuclear cells (BMMCs) isolated from patient-derived bone marrow samples as detailed in Section 3.1 were stained with fluorescent antibodies specific for CD38 plus fluorescent Annexin V (AnnV) in the presence of 7-AAD as a vital dye, as detailed in Section 3.4, followed by signal acquisition on flow cytometry and gating as detailed in Section 3.5. Representative dot plots and histograms from one sample are reported. Percentages indicate the numbers of event within specific gates.

2. As an alternative, Annexin V can be produced, purified and conjugated with DY-634 as detailed in Ref. Logue et al. (2009).
3. BM samples can be obtained by hematology hospital services or biobanks. Importantly, all studies handling human samples must be conducted according to the guidelines established by the Declaration of Helsinki (Hurst, 2014; Millum et al., 2013; Parsa-Parsi et al., 2013; World Medical, 2013). Additionally, written informed consent must be obtained from all subjects involved in the study.
4. 100 U/mL IL-6 (PeproTech, #200-06) must be added to the BMMC complete medium to carry out experiments determining the exposure of CALR or PS on the surface of $CD38^+$ plasma cells as driven by experimental interventions (during which plasma cells remain in culture). However, IL-6 is dispensable for direct *ex vivo* studies (when culture is not required).
5. PBS can also be prepared according to the following recipe: 137 mM NaCl, 10 mM Na_2HPO_4, and 1.8 mM KH_2PO_4, pH: 7.4.
6. Cell counting can also be performed on a common hemocytometer such as Neubauer or a Burker chamber.
7. The incubator is needed to quantify the exposure of CALR or PS on the surface of $CD38^+$ plasma cells as driven by experimental interventions, but not for direct *ex vivo* determinations (*see also Note 4*).
8. This is important to inactivate the complement system normally contained in FBS, which may otherwise lyse BMMCs (as well as other hematopoietic cells) (Ayuk et al., 2005; Celis & Celis, 1983; Zand et al., 2006).
9. Volume is normally limited (2–5 mL) as the clinical procedure to obtain samples is very invasive and samples are primary used for diagnostic procedures.
10. In the case that phases mix with each other, the experiment is compromised.
11. Electronic pipet-aids should be placed in drop-wise mode to minimize the chances of mixing phases.
12. Deactivating the brake is important to prevent separated phases to mix owing to an excessively harsh deceleration.
13. This is optional but recommended. Considerable care should be employed to avoid interfering with the BMMC layer.
14. Staining can also be performed in 1.5 mL microcentrifuge tubes but sample should next be transferred to 5 mL Round-Bottom Polypropylene Test Tubes for flow cytometry.

15. Protecting samples for direct light minimizes fluorochrome photobleaching, which may otherwise negatively impact on signal acquisition.
16. A reduced flow rate ensures sharper signal detection.
17. The percentage of CD38$^+$ plasma cells may be useful for additional analyses and/or experiment planning. A more accurate assessment of CD38$^+$ plasma cell abundance can be obtained directly staining 100 μL of the BMMC suspension with 8 μL APC or 4 μL FITC Mouse Anti-Human CD38 (HIT2) followed by flow cytometry. Only samples containing >0.1% CD38$^+$ plasma cells are suitable for additional experiments.
18. 7-AAD negativity is particularly important to ensure that surface-exposed (rather than total) CALR or PS is selectively detected.
19. Viability can also be conveniently assessed with propidium iodide (Miltenyi Biotec, #130-093-233), TO-PRO-3 (Invitrogen, #10710194) or DAPI (Thermo Scientific, #10116287), upon ensuring no spectral overlap with the antibody in use.
20. Mean Fluorescence Intensity (MFI) can also be employed to quantify CALR exposure upon normalization to MFI in the same sample stained with isotype control antibodies.
21. This is important as the binding between PS and Annexin V is Ca^{2+}-dependent (Blackwood & Ernst, 1990; Lemmon, 2008; Schlaepfer et al., 1987). If recombinant Annexin V is produced and conjugated with DY-634 in the laboratory, an adequate ABB can be obtained as per the following recipe: 10 mM HEPES/NaOH, 140 mM NaCl and 2.5 mM $CaCl_2$ in PBS, pH: 7.4.
22. While these volumes refer to a single sample, the staining solution should be prepared collectively for all samples included in the experiments (including 10% surplus) to homogenize staining conditions.
23. MFI is generally not employed to quantify PS exposure.

5. Concluding remarks

Stressed and dying cells expose a number of molecules on their surface, and this is paramount for the organism to properly react to cell death in support of systemic homeostasis (Freeman & Grinstein, 2021; Galluzzi et al., 2018; Poon & Ravichandran, 2024). In this context, CALR and PS mediate diametrically opposed functions, the former promoting phagocytosis by APC precursors, which potentially culminates with adaptive immune responses, the latter promoting efferocytosis by macrophages, which generally results in suppressed inflammation and immunity (Boada-Romero et al., 2020; Mehrotra & Ravichandran, 2022; Rothlin et al., 2021). Thus, assessing the propensity of malignant cells to spontaneously expose CALR or PS on their surface may not only provide insights into the overall degree of stress these cells are experiencing *in vivo*, but also convey prognostic and/or predictive information, as previously demonstrated in a number of patient cohorts (Fucikova et al., 2015). Along similar lines, testing the propensity of malignant cells to expose CALR or PS on their surface upon therapy (be it *in vivo* or *ex vivo*) may provide valuable information for the design of novel therapeutic strategies building on the induction of ICD (Bezu et al., 2015; Calvillo-Rodríguez et al., 2023; Ghiringhelli & Rébé, 2024; Mishchenko et al., 2024; Pan et al., 2024; Vacchelli et al., 2012).

The protocol detailed herein provides a simple method to specifically quantify CALR and PS levels on the surface of viable CD38$^+$ plasma cells by flow cytometry, which (with some variations reflecting tissue of origin and species) should be simply adaptable to other malignant cells from humans and rodents. To obtain an increasingly precise characterization of intratumoral heterogeneity in MM with respect to CALR and PS exposure, we ultimately aim at introducing key immunophenotypic MM markers in this analysis, notably, CD19, syndecan 1 (SDC1, best known as CD138), protein tyrosine phosphatase receptor type C (PTPRC, best known as CD45), neural cell adhesion molecule 1 (NCAM1, best known as CD56) and KIT proto-oncogene, receptor tyrosine kinase (KIT, also known as CD117 (Kumar et al., 2010; San Miguel et al., 2002). Despite these and other possible ameliorations, the flow cytometry-assisted quantification of CALR and PS on CD38$^+$ plasma cells stands out as a convenient tool to investigate spontaneous or therapy-driven immunogenic stress responses in BM samples from patients with MM.

Acknowledgments

M.B.-V. was supported by a predoctoral fellowship from Spanish Ministry of Universities (FPU17/02586) and A.S.-D.V. and N.J.-A. were also supported by a predoctoral fellowship from Gobierno de Aragón. The L.G. lab is/has been supported (as a PI unless otherwise indicated) by one NIH R01 grant (#CA271915), by two Breakthrough Level 2 grants from the US DoD BCRP (#BC180476P1, #BC210945), by a grant from the STARR Cancer Consortium (#I16-0064), by a Transformative Breast Cancer Consortium Grant from the US DoD BCRP (#W81XWH2120034, PI: Formenti), by a U54 grant from NIH/NCI (#CA274291, PI: Deasy, Formenti, Weichselbaum), by the 2019 Laura Ziskin Prize in Translational Research (#ZP-6177, PI: Formenti) from the

Stand Up to Cancer (SU2C), by a Mantle Cell Lymphoma Research Initiative (MCL-RI, PI: Chen-Kiang) grant from the Leukemia and Lymphoma Society (LLS), by a Rapid Response Grant from the Functional Genomics Initiative (New York, US), by a pre-SPORE grant (PI: Demaria, Formenti), a Collaborative Research Initiative Grant and a Clinical Trials Innovation Grant from the Sandra and Edward Meyer Cancer Center (New York, US), by startup funds from the Dept. of Radiation Oncology at Weill Cornell Medicine (New York, US), by industrial collaborations with Lytix Biopharma (Oslo, Norway), Promontory (New York, US) and Onxeo (Paris, France), as well as by donations from Promontory (New York, US), the Luke Heller TECPR2 Foundation (Boston, US), Sotio a.s. (Prague, Czech Republic), Lytix Biopharma (Oslo, Norway), Onxeo (Paris, France), Ricerchiamo (Brescia, Italy), and Noxopharm (Chatswood, Australia). The IM lab is/has been supported by grants #PID2019-105128RB-I00 and #PID2022-136799OB-I00 funded by MCIN/AEI/10.13039/501100011033 and grant #B31_20R funded by Gobierno de Aragón.

Conflicts of interests

L.G. is/has been holding research contracts with Lytix Biopharma, Promontory and Onxeo, has received consulting/advisory honoraria from Boehringer Ingelheim, AstraZeneca, AbbVie, OmniSEQ, Onxeo, The Longevity Labs, Inzen, Imvax, Sotio, Promontory, Noxopharm, EduCom, and the Luke Heller TECPR2 Foundation, and holds Promontory stock options. The other authors have no conflicts of interest to declare.

References

Adkins, I., et al. (2014). Physical modalities inducing immunogenic tumor cell death for cancer immunotherapy. *Oncoimmunology*, *3*, e968434. https://doi.org/10.4161/21624011.2014.968434.

Ayuk, F. A., et al. (2005). Antithymocyte globulin induces complement-dependent cell lysis and caspase-dependent apoptosis in myeloma cells. *Experimental Hematology*, *33*, 1531–1536. https://doi.org/10.1016/j.exphem.2005.08.004.

Bezu, L., et al. (2015). Combinatorial strategies for the induction of immunogenic cell death. *Frontiers in Immunology*, *6*, 187. https://doi.org/10.3389/fimmu.2015.00187.

Bezu, L., et al. (2018a). eIF2α phosphorylation: A hallmark of immunogenic cell death. *Oncoimmunology*, 7, e1431089. https://doi.org/10.1080/2162402x.2018.1431089.

Bezu, L., et al. (2018b). eIF2α phosphorylation is pathognomonic for immunogenic cell death. *Cell Death and Differentiation*, *25*, 1375–1393. https://doi.org/10.1038/s41418-017-0044-9.

Blackwood, R. A., & Ernst, J. D. (1990). Characterization of Ca2(+)-dependent phospholipid binding, vesicle aggregation and membrane fusion by annexins. *The Biochemical Journal*, *266*, 195–200. https://doi.org/10.1042/bj2660195.

Boada-Romero, E., et al. (2020). The clearance of dead cells by efferocytosis. *Nature Reviews. Molecular Cell Biology*, *21*, 398–414. https://doi.org/10.1038/s41580-020-0232-1.

Böttcher, J. P., et al. (2018). NK cells stimulate recruitment of cDC1 into the tumor microenvironment promoting cancer immune control. *Cell*, *172*, 1022–1037. e1014. https://doi.org/10.1016/j.cell.2018.01.004.

Calvillo-Rodríguez, K. M., et al. (2023). Immunotherapies inducing immunogenic cell death in cancer: Insight of the innate immune system. *Frontiers in Immunology*, *14*, 1294434. https://doi.org/10.3389/fimmu.2023.1294434.

Casares, N., et al. (2005). Caspase-dependent immunogenicity of doxorubicin-induced tumor cell death. *The Journal of Experimental Medicine*, *202*, 1691–1701. https://doi.org/10.1084/jem.20050915.

Celis, B., & Celis, E. (1983). Complement-mediated killing of myeloma tumour cells: Differences in susceptibility to lysis by antibodies and complement are independent of antigen expression and antibody binding. *Immunology*, *49*, 321–328.

Cytlak, U. M., et al. (2022). Immunomodulation by radiotherapy in tumour control and normal tissue toxicity. *Nature Reviews. Immunology*, *22*, 124–138. https://doi.org/10.1038/s41577-021-00568-1.

Czabotar, P. E., & Garcia-Saez, A. J. (2023). Mechanisms of BCL-2 family proteins in mitochondrial apoptosis. *Nature Reviews. Molecular Cell Biology*, *24*, 732–748. https://doi.org/10.1038/s41580-023-00629-4.

Diepstraten, S. T., et al. (2022). The manipulation of apoptosis for cancer therapy using BH3-mimetic drugs. *Nature Reviews. Cancer*, *22*, 45–64. https://doi.org/10.1038/s41568-021-00407-4.

Doran, A. C., et al. (2020). Efferocytosis in health and disease. *Nature Reviews. Immunology*, *20*, 254–267. https://doi.org/10.1038/s41577-019-0240-6.

Efimova, I., et al. (2020). Vaccination with early ferroptotic cancer cells induces efficient antitumor immunity. *Journal for Immunotherapy of Cancer*, *8*. https://doi.org/10.1136/jitc-2020-001369.

Fadok, V. A., et al. (1992). Exposure of phosphatidylserine on the surface of apoptotic lymphocytes triggers specific recognition and removal by macrophages. *Journal of Immunology*, *148*, 2207–2216.

Fadok, V. A., et al. (2000). A receptor for phosphatidylserine-specific clearance of apoptotic cells. *Nature*, *405*, 85–90. https://doi.org/10.1038/35011084.

Freeman, S., & Grinstein, S. (2021). Promoters and antagonists of phagocytosis: A plastic and tunable response. *Annual Review of Cell and Developmental Biology*, *37*, 89–114. https://doi.org/10.1146/annurev-cellbio-120219-055903.

Fuchs, Y., & Steller, H. (2015). Live to die another way: Modes of programmed cell death and the signals emanating from dying cells. *Nature Reviews. Molecular Cell Biology*, *16*, 329–344. https://doi.org/10.1038/nrm3999.

Fucikova, J., et al. (2015). Prognostic and predictive value of DAMPs and DAMP-associated processes in Cancer. *Frontiers in Immunology*, *6*, 402. https://doi.org/10.3389/fimmu.2015.00402.

Fucikova, J., et al. (2016a). Calreticulin expression in human non-small cell lung cancers correlates with increased accumulation of antitumor immune cells and favorable prognosis. *Cancer Research*, *76*, 1746–1756. https://doi.org/10.1158/0008-5472.Can-15-1142.

Fucikova, J., et al. (2016b). Calreticulin exposure by malignant blasts correlates with robust anticancer immunity and improved clinical outcome in AML patients. *Blood*, *128*, 3113–3124. https://doi.org/10.1182/blood-2016-08-731737.

Fucikova, J., et al. (2021). Calreticulin and cancer. *Cell Research*, *31*, 5–16. https://doi.org/10.1038/s41422-020-0383-9.

Galluzzi, L., et al. (2018). Linking cellular stress responses to systemic homeostasis. *Nature Reviews. Molecular Cell Biology*, *19*, 731–745. https://doi.org/10.1038/s41580-018-0068-0.

Galluzzi, L., et al. (2020a). Consensus guidelines for the definition, detection and interpretation of immunogenic cell death. *Journal for Immunotherapy of Cancer*, *8*. https://doi.org/10.1136/jitc-2019-000337.

Galluzzi, L., et al. (2020b). Immunostimulation with chemotherapy in the era of immune checkpoint inhibitors. *Nature Reviews. Clinical Oncology*, *17*, 725–741. https://doi.org/10.1038/s41571-020-0413-z.

Galluzzi, L., et al. (2023). Immunogenic cell death in cancer: Concept and therapeutic implications. *Journal of Translational Medicine*, *21*, 162. https://doi.org/10.1186/s12967-023-04017-6.

Gardai, S. J., et al. (2005). Cell-surface calreticulin initiates clearance of viable or apoptotic cells through trans-activation of LRP on the phagocyte. *Cell*, *123*, 321–334. https://doi.org/10.1016/j.cell.2005.08.032.

Garg, A. D., et al. (2016). Dendritic cell vaccines based on immunogenic cell death elicit danger signals and T cell-driven rejection of high-grade glioma. *Science Translational Medicine*, *8*. https://doi.org/10.1126/scitranslmed.aae0105. 328ra327.

Ghiringhelli, F., & Rébé, C. (2024). Using immunogenic cell death to improve anticancer efficacy of immune checkpoint inhibitors: From basic science to clinical application. *Immunological Reviews*, *321*, 335–349. https://doi.org/10.1111/imr.13263.

Giampazolias, E., et al. (2017). Mitochondrial permeabilization engages NF-κB-dependent anti-tumour activity under caspase deficiency. *Nature Cell Biology*, *19*, 1116–1129. https://doi.org/10.1038/ncb3596.

Gong, T., et al. (2020). DAMP-sensing receptors in sterile inflammation and inflammatory diseases. *Nature Reviews. Immunology*, *20*, 95–112. https://doi.org/10.1038/s41577-019-0215-7.

Hadian, K., & Stockwell, B. R. (2023). The therapeutic potential of targeting regulated non-apoptotic cell death. *Nature Reviews. Drug Discovery*, *22*, 723–742. https://doi.org/10.1038/s41573-023-00749-8.

Han, C., et al. (2020). Tumor cells suppress radiation-induced immunity by hijacking caspase 9 signaling. *Nature Immunology*, *21*, 546–554. https://doi.org/10.1038/s41590-020-0641-5.

Harrington, K., et al. (2019). Optimizing oncolytic virotherapy in cancer treatment. *Nature Reviews. Drug Discovery*, *18*, 689–706. https://doi.org/10.1038/s41573-019-0029-0.

Hsu, W. M., et al. (2005). Calreticulin expression in neuroblastoma—A novel independent prognostic factor. *Annals of Oncology*, *16*, 314–321. https://doi.org/10.1093/annonc/mdi062.

Huang, Q., et al. (2011). Caspase 3-mediated stimulation of tumor cell repopulation during cancer radiotherapy. *Nature Medicine*, *17*, 860–866. https://doi.org/10.1038/nm.2385.

Hurst, S. A. (2014). Declaration of Helsinki and protection for vulnerable research participants. *JAMA*, *311*, 1252. https://doi.org/10.1001/jama.2014.1272.

Jhunjhunwala, S., et al. (2021). Antigen presentation in cancer: Insights into tumour immunogenicity and immune evasion. *Nature Reviews. Cancer*, *21*, 298–312. https://doi.org/10.1038/s41568-021-00339-z.

Kasikova, L., et al. (2019). Calreticulin exposure correlates with robust adaptive antitumor immunity and favorable prognosis in ovarian carcinoma patients. *Journal for Immunotherapy of Cancer*, 7, 312. https://doi.org/10.1186/s40425-019-0781-z.

Kepp, O., et al. (2020). Oncolysis without viruses—Inducing systemic anticancer immune responses with local therapies. *Nature Reviews. Clinical Oncology*, *17*, 49–64. https://doi.org/10.1038/s41571-019-0272-7.

Kroemer, G., et al. (2022). Immunogenic cell stress and death. *Nature Immunology*, *23*, 487–500. https://doi.org/10.1038/s41590-022-01132-2.

Krysko, D. V., et al. (2012). Immunogenic cell death and DAMPs in cancer therapy. *Nature Reviews. Cancer*, *12*, 860–875. https://doi.org/10.1038/nrc3380.

Kumar, S., et al. (2010). Immunophenotyping in multiple myeloma and related plasma cell disorders. *Best Practice & Research. Clinical Haematology*, *23*, 433–451. https://doi.org/10.1016/j.beha.2010.09.002.

Lang, F., et al. (2022). Identification of neoantigens for individualized therapeutic cancer vaccines. *Nature Reviews. Drug Discovery*, *21*, 261–282. https://doi.org/10.1038/s41573-021-00387-y.

Lei, G., et al. (2022). Targeting ferroptosis as a vulnerability in cancer. *Nature Reviews. Cancer*, *22*, 381–396. https://doi.org/10.1038/s41568-022-00459-0.

Lemmon, M. A. (2008). Membrane recognition by phospholipid-binding domains. *Nature Reviews. Molecular Cell Biology*, *9*, 99–111. https://doi.org/10.1038/nrm2328.

Li, M. O., et al. (2003). Phosphatidylserine receptor is required for clearance of apoptotic cells. *Science, 302*, 1560–1563. https://doi.org/10.1126/science.1087621.

Logue, S. E., et al. (2009). Expression, purification and use of recombinant annexin V for the detection of apoptotic cells. *Nature Protocols, 4*, 1383–1395. https://doi.org/10.1038/nprot.2009.143.

Matzinger, P. (2002). The danger model: A renewed sense of self. *Science, 296*, 301–305. https://doi.org/10.1126/science.1071059.

McLaughlin, M., et al. (2020). Inflammatory microenvironment remodelling by tumour cells after radiotherapy. *Nature Reviews. Cancer, 20*, 203–217. https://doi.org/10.1038/s41568-020-0246-1.

Mehrotra, P., & Ravichandran, K. S. (2022). Drugging the efferocytosis process: Concepts and opportunities. *Nature Reviews. Drug Discovery, 21*, 601–620. https://doi.org/10.1038/s41573-022-00470-y.

Mellman, I., et al. (2023). The cancer-immunity cycle: Indication, genotype, and immunotype. *Immunity, 56*, 2188–2205. https://doi.org/10.1016/j.immuni.2023.09.011.

Millum, J., et al. (2013). The 50th anniversary of the declaration of Helsinki: Progress but many remaining challenges. *JAMA, 310*, 2143–2144. https://doi.org/10.1001/jama.2013.281632.

Mishchenko, T. A., et al. (2024). Targeting immunogenic cell death for glioma immunotherapy. *Trends in Cancer, 10*, 8–11. https://doi.org/10.1016/j.trecan.2023.10.005.

Ning, X., et al. (2019). Apoptotic caspases suppress type I interferon production via the cleavage of cGAS, MAVS, and IRF3. *Molecular Cell, 74*, 19–31. e17 https://doi.org/10.1016/j.molcel.2019.02.013.

Obeid, M., et al. (2007). Leveraging the immune system during chemotherapy: Moving calreticulin to the cell surface converts apoptotic death from "silent" to immunogenic. *Cancer Research, 67*, 7941–7944. https://doi.org/10.1158/0008-5472.Can-07-1622.

Ogden, C. A., et al. (2001). C1q and mannose binding lectin engagement of cell surface calreticulin and CD91 initiates macropinocytosis and uptake of apoptotic cells. *The Journal of Experimental Medicine, 194*, 781–795. https://doi.org/10.1084/jem.194.6.781.

Palanivelu, L., et al. (2022). Immunogenic cell death: The cornerstone of oncolytic viro-immunotherapy. *Frontiers in Immunology, 13*, 1038226. https://doi.org/10.3389/fimmu.2022.1038226.

Pan, H., et al. (2024). Preconditioning with immunogenic cell death-inducing treatments for subsequent immunotherapy. *International Review of Cell and Molecular Biology, 382*, 279–294. https://doi.org/10.1016/bs.ircmb.2023.06.001.

Panaretakis, T., et al. (2009). Mechanisms of pre-apoptotic calreticulin exposure in immunogenic cell death. *The EMBO Journal, 28*, 578–590. https://doi.org/10.1038/emboj.2009.1.

Parsa-Parsi, R., et al. (2013). Reconsidering the declaration of Helsinki. *Lancet, 382*, 1246–1247. https://doi.org/10.1016/S0140-6736(13)62094-2.

Pawaria, S., & Binder, R. J. (2011). CD91-dependent programming of T-helper cell responses following heat shock protein immunization. *Nature Communications, 2*, 521. https://doi.org/10.1038/ncomms1524.

Peng, R. Q., et al. (2010). Expression of calreticulin is associated with infiltration of T-cells in stage IIIB colon cancer. *World Journal of Gastroenterology, 16*, 2428–2434. https://doi.org/10.3748/wjg.v16.i19.2428.

Poon, I. K. H., & Ravichandran, K. S. (2024). Targeting efferocytosis in inflammaging. *Annual Review of Pharmacology and Toxicology, 64*, 339–357. https://doi.org/10.1146/annurev-pharmtox-032723-110507.

Rodriguez-Ruiz, M. E., et al. (2019). Apoptotic caspases inhibit abscopal responses to radiation and identify a new prognostic biomarker for breast cancer patients. *Oncoimmunology, 8*, e1655964. https://doi.org/10.1080/2162402x.2019.1655964.

Rodriguez-Ruiz, M. E., et al. (2020). Immunological impact of cell death signaling driven by radiation on the tumor microenvironment. *Nature Immunology*, *21*, 120–134. https://doi.org/10.1038/s41590-019-0561-4.

Rongvaux, A., et al. (2014). Apoptotic caspases prevent the induction of type I interferons by mitochondrial DNA. *Cell*, *159*, 1563–1577. https://doi.org/10.1016/j.cell.2014.11.037.

Rothlin, C. V., et al. (2021). Determining the effector response to cell death. *Nature Reviews. Immunology*, *21*, 292–304. https://doi.org/10.1038/s41577-020-00456-0.

San Miguel, J. F., et al. (2002). Immunophenotypic evaluation of the plasma cell compartment in multiple myeloma: A tool for comparing the efficacy of different treatment strategies and predicting outcome. *Blood*, *99*, 1853–1856. https://doi.org/10.1182/blood.v99.5.1853.

Schlaepfer, D. D., et al. (1987). Structural and functional characterization of endonexin II, a calcium- and phospholipid-binding protein. *Proceedings of the National Academy of Sciences of the United States of America*, *84*, 6078–6082. https://doi.org/10.1073/pnas.84.17.6078.

Serrano Del Valle, A., et al. (2022). Ecto-calreticulin expression in multiple myeloma correlates with a failed anti-tumoral immune response and bad prognosis. *Oncoimmunology*, *11*, 2141973. https://doi.org/10.1080/2162402x.2022.2141973.

Shalhout, S. Z., et al. (2023). Therapy with oncolytic viruses: Progress and challenges. *Nature Reviews. Clinical Oncology*, *20*, 160–177. https://doi.org/10.1038/s41571-022-00719-w.

Sprooten, J., et al. (2023). Trial watch: Chemotherapy-induced immunogenic cell death in oncology. *Oncoimmunology*, *12*, 2219591. https://doi.org/10.1080/2162402x.2023.2219591.

Stern, L. J., et al. (2024). Non-mutational neoantigens in disease. *Nature Immunology*, *25*, 29–40. https://doi.org/10.1038/s41590-023-01664-1.

Tatsuno, K., et al. (2019). Extracorporeal photochemotherapy induces bona fide immunogenic cell death. *Cell Death & Disease*, *10*, 578. https://doi.org/10.1038/s41419-019-1819-3.

Truxova, I., et al. (2020). Calreticulin exposure on malignant blasts correlates with improved natural killer cell-mediated cytotoxicity in acute myeloid leukemia patients. *Haematologica*, *105*, 1868–1878. https://doi.org/10.3324/haematol.2019.223933.

Vacchelli, E., et al. (2012). Trial watch: Chemotherapy with immunogenic cell death inducers. *Oncoimmunology*, *1*, 179–188. https://doi.org/10.4161/onci.1.2.19026.

Vandivier, R. W., et al. (2002). Role of surfactant proteins a, D, and C1q in the clearance of apoptotic cells in vivo and in vitro: Calreticulin and CD91 as a common collectin receptor complex. *Journal of Immunology*, *169*, 3978–3986. https://doi.org/10.4049/jimmunol.169.7.3978.

Ventura, A., et al. (2018). Extracorporeal photochemotherapy drives monocyte-to-dendritic cell maturation to induce anticancer immunity. *Cancer Research*, *78*, 4045–4058. https://doi.org/10.1158/0008-5472.Can-18-0171.

Vitale, I., et al. (2023). Apoptotic cell death in disease—Current understanding of the NCCD 2023. *Cell Death and Differentiation*, *30*, 1097–1154. https://doi.org/10.1038/s41418-023-01153-w.

Wemeau, M., et al. (2010). Calreticulin exposure on malignant blasts predicts a cellular anticancer immune response in patients with acute myeloid leukemia. *Cell Death & Disease*, *1*, e104. https://doi.org/10.1038/cddis.2010.82.

White, M. J., et al. (2014). Apoptotic caspases suppress mtDNA-induced STING-mediated type I IFN production. *Cell*, *159*, 1549–1562. https://doi.org/10.1016/j.cell.2014.11.036.

World Medical A. (2013). World medical association declaration of Helsinki: Ethical principles for medical research involving human subjects. *JAMA*, *310*, 2191–2194. https://doi.org/10.1001/jama.2013.281053.

Xu, Q., et al. (2018). Significance of calreticulin as a prognostic factor in endometrial cancer. *Oncology Letters*, *15*, 8999–9008. https://doi.org/10.3892/ol.2018.8495.

Yamazaki, T., et al. (2020). Mitochondrial DNA drives abscopal responses to radiation that are inhibited by autophagy. *Nature Immunology*, *21*, 1160–1171. https://doi.org/10.1038/s41590-020-0751-0.

Yatim, N., et al. (2015). RIPK1 and NF-κB signaling in dying cells determines cross-priming of $CD8^+$ T cells. *Science*, *350*, 328–334. https://doi.org/10.1126/science.aad0395.

Yuan, J., & Ofengeim, D. (2023). A guide to cell death pathways. *Nature Reviews. Molecular Cell Biology*. https://doi.org/10.1038/s41580-023-00689-6.

Zand, M. S., et al. (2006). Apoptosis and complement-mediated lysis of myeloma cells by polyclonal rabbit antithymocyte globulin. *Blood*, *107*, 2895–2903. https://doi.org/10.1182/blood-2005-06-2269.

Zelenay, S., et al. (2015). Cyclooxygenase-dependent tumor growth through evasion of immunity. *Cell*, *162*, 1257–1270. https://doi.org/10.1016/j.cell.2015.08.015.

9780443296222